药学前沿系列丛书

Stem Cell and Regenerative Medicine

干细胞与再生医学

主　编　程　芳

副主编　陈红波

·广州·

图书在版编目（CIP）数据

干细胞与再生医学：英文、汉文/程芳主编；陈红波副主编. —广州：中山大学出版社，2022. 10

（药学前沿系列丛书）

ISBN 978－7－306－07541－3

Ⅰ. ①干…　Ⅱ. ①程…　②陈…　Ⅲ. ①干细胞—细胞生物学—英、汉　②细胞—再生—生物医学工程—英、汉　Ⅳ. ①Q24　②R318

中国版本图书馆 CIP 数据核字（2022）第 084490 号

GANXIBAO YU ZAISHENGYIXUE

出 版 人：王天琪

策划编辑：陈　慧　鲁佳慧

责任编辑：鲁佳慧　罗永梅

封面设计：曾　斌

责任校对：李昭莹

责任技编：靳晓虹

出版发行：中山大学出版社

电　　话：编辑部 020－84110283，84113349，84111997，84110779，84110776

　　　　　发行部 020－84111998，84111981，84111160

地　　址：广州市新港西路 135 号

邮　　编：510275　　传　　真：020－84036565

网　　址：http：//www. zsup. com. cn　E-mail：zdcbs@ mail. sysu. edu. cn

印 刷 者：广东虎彩云印刷有限公司

规　　格：787mm×1092mm　1/16　12.125 印张　300 千字

版次印次：2022 年 10 月第 1 版　2024 年 11 月第 2 次印刷

定　　价：59. 80 元

本书编委会

主　编： 程　芳

副主编： 陈红波

编　委： 张伟娴　曾宇晴　苏丹丹　黄如凡

韩雨航　石培霖

编写团队： 中山大学药学院（深圳）

Preface
前　言

干细胞与再生医学作为当前生命科学领域的发展前沿之一，具有重大临床应用价值。干细胞领域是当今生物医学最热门的领域之一，在治疗传统医学十分棘手的病证上已显示出惊人的潜力，例如，在新冠病毒肺炎疫情中，利用间充质干细胞移植挽救危重症感染患者屡屡成功。不同于传统医疗方式，多重策略的干细胞再生组织器官疗法将改变传统医学对于坏死性和损伤性等疾病的治疗手段，为疾病的机理研究和临床运用带来革命性变化。

干细胞是高等多细胞生物体内一种具有自我更新及多向分化潜能的未分化或低分化的细胞。干细胞除了具备自我更新与多向分化潜能外，还具有归巢性，即被重新注入人体的干细胞会自动靶向至体内受损部位，从而达到快速修复组织的目的。因此，干细胞作为一种新型医疗策略，可能颠覆常规传统治疗手段，在治疗糖尿病、癌症、心脑血管疾病、肝脏疾病、神经元损伤疾病等多种重大疾病中都有着极大潜能，被称为医学界的“万用细胞”。除此之外，干细胞还可用于疾病模型研究、药物开发与筛选、再生医学、精准医学及基因治疗等，用途广泛，其研究与应用前景一片光明。

干细胞疗法，即再生医疗技术，是将体外健康的干细胞或干细胞外泌体，通过特殊的移植技术移植到患者体内目标组织区域，在体内分化、再生成新的细胞或组织，从而弥补组织细胞的衰老、损伤与死亡，发挥组织修复功能；或在体内通过分泌各种蛋白质与因子来发挥免疫调节功能，从而恢复机体功能。

干细胞治疗相关研究在近 30 年来一直是生命科学领域中的研究热点，其在临床疾病诊治与药物应用方面呈现出极大的发展前景与潜力。尤其随着细胞重编程技术与基因编辑技术的进步，成体干细胞与多能干细胞也将进一步促进干细胞疗法的发展。其中诱导多能干细胞几乎可以分化为所有细胞类型，且像胚胎干细胞一样具有强大的生长能力，因此在药物研发和临床应用中都具有巨大的潜力。患者特异性的诱导多能干细胞可用作疾病模型以进行进一步的病理学研究。

近年来，干细胞与再生医学领域的国际竞争日趋激烈。干细胞与再生医学在临床上的广泛研究与应用代表着未来医学发展的一大方向。尽管如此，目前干细胞技术在临床应用中仍存在安全隐患。哈佛大学的著名干细胞生物学家 Douglas Melton 教授曾经说过：“干细胞可能是承诺得最多但兑现得最少的领域。”之所以这么说，是因为干细胞技术在数十年基础研究阶段经历了大起大落，被各种造假丑闻缠身。因此，在展望干细胞技术与再生医学领域发展前景的同时，我们仍须重视人类干细胞研究带来的伦理争论，以及干细胞技术存在的潜在风险与障碍。唯有牢记医学之除病救民之根本，我们才能正确地运用好科学技术这把“双刃剑”。

本书主要介绍了干细胞与再生医学的相关研究进展，包括干细胞的基础研究与衍生研究，及其在不同生理和病理水平中的治疗作用。第一章主要介绍了干细胞的类型、来源、应用和目前面临的挑战。第二章以胚胎干细胞、间充质干细胞、诱导多能干细胞和成体神经干细胞为例，介绍了干细胞的分离与鉴定策略及实验方法。第三章则聚焦于干细胞与人类疾病间的密切联系，阐述了目前干细胞在神经系统疾病、心脏疾病、糖尿病、骨关节炎和伤口愈合等方面的治疗策略与研究进展。第四章介绍了干细胞微环境，以及再生医学中通过靶向干细胞微环境来治疗疾病的组织工程。第五章讨论了肿瘤干细胞学说，介绍了肿瘤干细胞的基本概念与特性、耐药机制、微环境及目前研究中遇到的挑战。第六章与第七章介绍了药物和疾病的新型研究平台与工具——类器官，阐述了其在癌症医学、精准医学、再生医学、药物开发（筛选和毒理学）、疾病模型和基础研究中的研究进展与应用，理性探讨其局限性；同时介绍了类器官的不同类型，着重介绍了患者来源的人类肝癌类器官的研究进展。第八章介绍了近年来研究热度极高的细胞外囊泡，作为干细胞治疗的潜在替代品，我们对这种可作为诊断、预后和治疗性工具的生物标记物的发生、分类、组成、功能与应用等研究的进展进行了总结。第九章介绍了目前生命科学领域中作为新型生物技术疗法的基因治疗及有望显著提高基因治疗的持久性与安全性的基因编辑技术，以期为疾病治疗策略提供更多新的思路。

我们希望本书能帮助读者更好地了解干细胞与再生医学领域的基础知识与研究现状，为该领域的学习和研究提供更多思考路径，从而在面对挑战时不至望而却步。我们应该坚信，没有什么科学问题是无法解决的，科研工作者们只是需要更多的时间与更多的努力。

最后，我要感谢中山大学出版社及参与本书编写与出版的所有编委和编辑老师们。此外，特向在本教程中引用和参考的已注明和未注明的教材、专著、报刊的作者表示诚挚的谢意。

本教程虽经多次修改，但由于编者能力有限，不足之处在所难免，敬请专家和读者批评指正。

编者

2022 年 6 月

Contents
目 录

Chapter 6 Organoids

Chapter 7 Different Types of Organoids

Chapter 8 Extracellular Vesicles

Chapter 9 Genome Editing and Gene Therapy

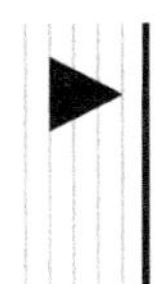

Chapter 1 Introduction to Stem Cell Biology
第一章 干细胞生物学概述

[中文导读]

干细胞与再生医学（stem cells and regenerative medicine）代表了现代生命科学发展的前沿，是现代临床医学的一种崭新的治疗模式，对医学的发展具有引领作用。干细胞是了解人体器官发生和持续再生能力的重要工具，其具有自我更新、多向分化、免疫调节和归巢等特点，为开发用于研究新药药理学的生物学模型提供了可能性。干细胞疗法，又称再生医疗技术，是把健康的干细胞移植到患者体内，以修复或替换受损细胞或组织，从而达到治愈的目的。利用干细胞可以实现真正意义上的个体化治疗。

再生医学标志着医学将步入重建、再生、“制造”、替代组织器官的新时代。基于干细胞修复与再生能力的再生医学有望成为继药物治疗、手术治疗后第三种治疗途径，前景广阔，能为人类面临的大多数医学难题带来新的希望，如心血管疾病、自身免疫性疾病、糖尿病、恶性肿瘤、阿尔茨海默病、帕金森病、先天性遗传缺陷等疾病和各种组织器官损伤的治疗；其相关基础与应用研究和现代生物医学技术的结合，也将使人类修复和制造组织器官的梦想得以实现，对医学治疗和再生理论的发展有重大影响。

迄今为止，人类已经开展了大量关于干细胞的临床前研究和临床试验，一些临床研究报告了令人鼓舞的基于细胞医学的新治疗策略的发展结果。同时，社会必须严格监管开拓性科学研究，以确保它们既符合伦理道德又具有安全性。目前，在使用以干细胞为主的医学解决方案时，仍存在许多风险和障碍，因此，还须进一步地探索与完善。

1 Stem Cell Research Timeline—Highlights

1.1 Stem Cell Research History

In 1961, Canadian scientists Till and McCulloch discovered hematopoietic stem cells from the bone marrow and first proposed the concept of pluripotent stem cells (PSCs)[1]. In 1968, the University of Washington completed the world's first bone marrow transplant, which is intravenous injection of successfully matched bone marrow-derived hematopoietic stem cells into leukemia patients. Seeing the development prospects of stem cell research, after 1990, countries around the world strongly supported stem cell research. In 1996, British scientists cloned a sheep named Dolly using mammary stem cells and tissue engineering technology[2], and a new era of stem cell begun.

In 1998, American scientists isolated an embryonic stem cell line, human pluripotent stem cells, from human blastocysts[3]. Since then, the development of stem cells has entered the stage of clinical research. After 2000, stem cells have been widely used in clinical research on diabetes, nerve damage, lupus erythematosus, and liver disease. In 2003, scientists extracted dental pulp mesenchymal stem cells from children's deciduous teeth. In 2005, Dvorak revealed that an involvement of autocrine fibroblast growth factor (FGF) signals in the maintenance of proliferating human embryonic stem cells (hESCs) in the undifferentiated state.

In 2006 and 2007, the team of Japanese scientist Shinya Yamanaka used viral vectors to transfer four transcription factors (Oct4, Sox2, Klf4, c-Myc) into differentiated fibroblasts, and reprogrammed fibroblasts to obtain a new type of pluripotent stem cells, i. e. induced pluripotent stem cells (iPSCs). The discovery of iPSCs has aroused strong repercussions in the fields of stem cells, epigenetics, and biomedicine, further narrowing the distance between stem cells and clinical disease treatment[4]. And Nobel Prize for physiology or medicine was awarded to Shinya Yamanaka for iPSCs in 2012.

To reduce the risk of tumorigenicity of viral integration into the host genome, Keisuke Okita reported the generation of human iPSCs using non-integrating plasmids in 2008. In 2009, Kaji and Woltjen reported self-excising vectors, e. g. piggy Bac transposon, can reprograms fibroblasts to human iPSCs. In 2010, Warren Luigi reported a simple, nonintegrating strategy for reprogramming cell fate based on administration of synthetic mRNA modified to overcome innate antiviral responses. In the same time, the first clinical trial using ESC-derived OPC1s in spinal cord injury begun by Geron, and GRNOPC1 was approved by FDA to enter early human clinical trials.

Cre-loxP excision generating nonintegrating human iPSCs was reported in 2011. In 2013, mouse fibroblasts were induced into iPSCs by using small molecular compound (e. g. valproic acid, CHIR99021, and FSK). In 2014, quiescent adult neural stem cells (NSCs) were identified, isolated and revealed important features of quiescent NSCs[5]. Multiparameter immunophenotypes and clinical applications of human adult stem cells from different sources have been extensively studied [6]. Adult mammalian NSCs is also a hot topic in the stem cell field [7]. In addition to Dolly, scientists successfully cloned macaque monkeys using fetal fibroblasts by somatic cell nuclear transfer in 2018 [8].

In 2015, bioprinting of human iPSCs was developed and these iPSCs can maintain their pluripotency or directing their differentiation into specific lineages. In 2016, CRISPR technology was used in iPSCs for the investigation of the molecular and cellular mechanisms underlying inherited diseases. Based on years of fundamental researches and pre-clinical studies, finally in 2017, a clinical trial of iPSC-derived retinal cells transplantation in woman suffering from advanced macular degeneration was conducted.

Over the past few decades, different key areas including sources of stem cells, next-generation *in vivo* reprogramming technology, induction of pluripotency with genomic modifications or chemical molecules, biomaterials, 3D cell culture and bioprinting have facilitated the research and development of pluripotent stem cells[9].

The history of stem cell is shown in Fig. 1 - 1[1-9].

1.2 The Turning Point in Stem Cell Therapy

The turning point in stem cell therapy appeared in 2006, when scientist Shinya Yamanaka, together with Kazutoshi Takahashi, discovered that it is possible to reprogram multipotent adult stem cells to the pluripotent state. This process avoided endangering the foetus' life.

Retrovirus-mediated transduction of mouse fibroblasts with four transcription factors (Oct-3/4, Sox2, Klf4, and c-Myc) that are mainly expressed in embryonic stem cells (ESCs) could induce the fibroblasts to become pluripotent. This new form of stem cells was named as iPSCs. One year later, the experiment also succeeded with human cells. After this success, the method opened a new field in stem cell research with a generation of iPSC lines that can be customized and biocompatible to the patient. Recently, studies have focused on reducing carcinogenesis and improving the conduction system.

The turning point was influenced by former discoveries that happened in 1962 and 1987. The former discovery was about scientist John Gurdon cloning frogs successfully by transferring a nucleus from a frog's somatic cells into an oocyte. This caused a complete reversion of somatic cell development. The results of Gurdon's experiment has become an immense discovery since it was previously believed that cell differentiation is a one-way street only, but his experiment suggested the opposite and demonstrated that it is even possible for a somatic cell to again acquire pluripotency[10-11].

The latter was a discovery made by R. L. Davis that focused on fibroblast DNA subtraction[12]. Three genes were found to be originally appeared in myoblasts. The enforced expression of only one of the genes, named myogenic differentiation 1 (Myod1), caused the conversion of fibroblasts into myoblasts, showing that reprogramming cells is possible, and it can even be used to transform cells from one lineage to another.

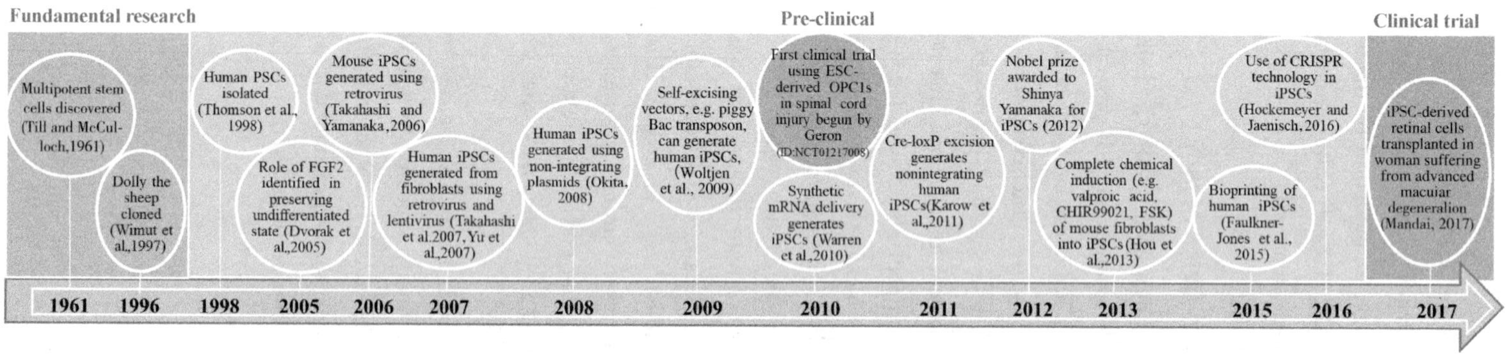

Fundamental research
Pre-clinical
Clinical trial

1961 Multipotent stem cells discovered (Till and McCul-loch,1961)
1996 Dolly the sheep cloned (Wimut et al.,1997)
1998 Human PSCs isolated (Thomson et al, 1998)
2005 Role of FGF2 identified in preserving undifferentiated state (Dvorak et al.,2005)
2006 Mouse iPSCs generated using retrovirus (Takahashi and Yamanaka,2006)
2007 Human iPSCs generated from fibroblasts using retrovirus and lentivirus (Takahashi et al.2007, Yu et al.,2007)
2008 Human iPSCs generated using non-integrating plasmids (Okita, 2008)
2009 Self-excising vectors, e.g. piggy Bac transposon, can generate human iPSCs, (Woltjen et al., 2009)
2010 First clinical trial using ESC-derived OPC1s in spinal cord injury begun by Geron (ID:NCT01217008)
2010 Synthetic mRNA delivery generates iPSCs (Warren et al.,2010)
2011 Cre-loxP excision generates nonintegrating human iPSCs(Karow et al.,2011)
2012 Nobel prize awarded to Shinya Yamanaka for iPSCs (2012)
2013 Complete chemical induction (e.g. valproic acid, CHIR99021, FSK) of mouse fibroblasts into iPSCs (Hou et al.,2013)
2015 Bioprinting of human iPSCs (Faulkner-Jones et al., 2015)
2016 Use of CRISPR technology in iPSCs (Hockemeyer and Jaenisch,2016)
2017 iPSC-derived retinal cells transplanted in woman suffering from advanced macuiar degeneralion (Mandai, 2017)

Fig.1-1 The timeline of major scientific advances during the history of stem cell research

Red shading represents fundamental research, yellow shading represents preclinical work, and green shading represents clinical trials.

Illustrated by 曾宇晴.

Reference:
LIU G, DAVID B T, TRAWCZYNSKI M, et al. Advances in pluripotent stem cells: history, mechanisms, technologies, and applications[J]. Stem cell reviews and reports. 2020，16(1)：3–32.

2 Introduction of Regenerative Medicine and Stem Cells

Regenerative medicine (RM) is an emerging discipline that studies how to restore tissue, organ defects, and physiological repair function loss caused by trauma and disease to normal morphology and function. RM mainly studies the mechanisms of stem cell proliferation, migration, and differentiation. The aim is to find a way to promote the body's self-repair and regeneration, and finally can build new tissues and organs to maintain, repair, regenerate or improve damaged tissues and organ functions. Regenerative medicine is a very comprehensive and interdisciplinary subject. At present, it is mainly applied to repair and replace damaged, diseased and defective tissues and organs by implanting human stem cells, tissues and organs, so as to restore their functions and structural reconstruction, and finally achieve the purpose of regeneration.

Stem cells, as regenerated seed cells, involve almost all fields of regenerative medicine. Stem cells are undifferentiated or poorly differentiated cells with self-renewal and multi-differentiation potential. Thousands of cells exist in almost any tissue or organ, from embryos to adults. Stem cells are the source of tissue and organ regeneration and the most important prerequisite for research and practice of regenerative medicine[13]. Stem cells have the following characteristics (Fig. 1 -2).

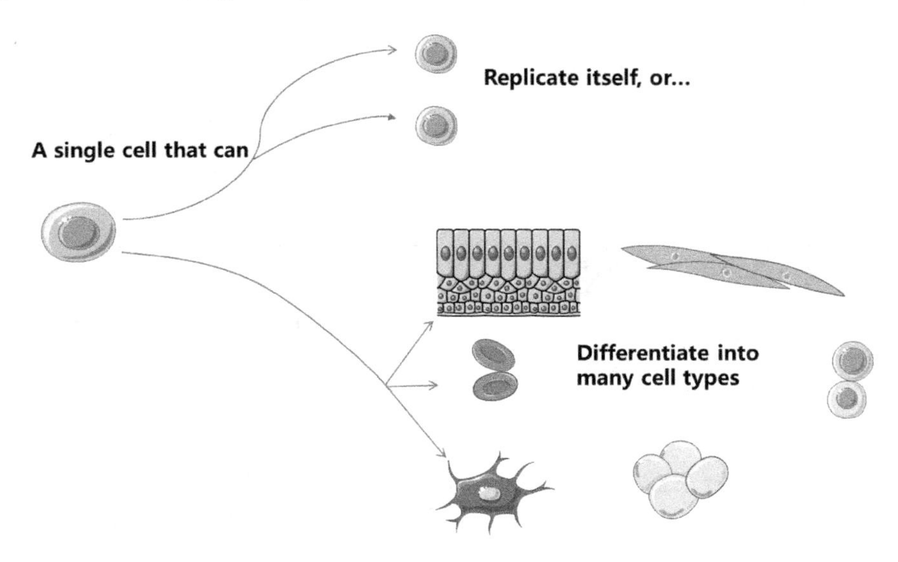

Fig. 1 -2 What is a stem cell?

Illustrated by 曾宇晴.

Reference:

LIN H T, OTSU M, NAKAUCHI H. Stem cell therapy: an exercise in patience and prudence [J]. Philosophical transactions of the Royal Society of London Series B, Biological sciences. 2013, 368 (1609): 20110334.

(1) Not itself terminally differentiated. "Stemness": the function of both self-renewal and differentiation.

(2) Unlimited division capacity. "Self-renewal": the ability to go through repeated cell divisions to produce cells identical to the parent cell.

(3) Daughter cells can choose either one.

(4) Stay a stem cell.

(5) Commit to terminally differentiate.

Therefore, it is best to understand the mechanism of stem cells and regeneration from the cellular and molecular level. As an important means and research core of regenerative medicine, thousands of cells cover many fields of basic and clinical medicine. In terms of basic research, the understanding of the molecular regulatory mechanisms of stem cell growth, migration, and differentiation is helpful to understand the basic life laws such as organ formation, repair, and functional reconstruction, and to study the mechanisms of regeneration and the methods of promoting regeneration.

3 Types and Origin of Stem Cell

3.1 Stem Cell Classification Based on Differentiation Potential

The ability to differentiate, one of the two main characteristics of stem cells, varies among stem cells depending on their origin and their derivation. All stem cells can be categorized according to their differentiation potential into 5 groups: totipotent or omnipotent, pluripotent, multipotent, oligopotent, and unipotent (Tab 1 – 1).

Tab. 1 – 1 Comparing stem cells' characteristics between ESCs, iPSCs and somatic stem cells (SSCs)

Characteristics	ESCs	iPSCs	SSCs
In vitro proliferation	very high	very high	limited
Potency	pluripotent	pluripotent	multipotent or restricted
Clinical application	induction necessary	induction necessary	directly transplantable
Tumorgenicity	possible	possible	none
Ethical issue	yes	yes	no

3.1.1 Totipotent Stem Cells

Totipotent or omnipotent stem cells are the most undifferentiated cells and are found in early development. A fertilized oocyte and the cells of the first two divisions are totipotent stem cells, as they differentiate into both embryonic and extra-embryonic tissues, thereby

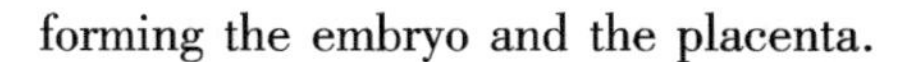

forming the embryo and the placenta.

3.1.2 Pluripotent Stem Cells

Pluripotent stem cells are able to differentiate into cells that arise from the 3 germ layers—ectoderm, endoderm, and mesoderm—from which all tissues and organs develop. ESCs were first derived from the inner cell mass (ICM) of the blastocyst. These cells are called induced pluripotent stem cells (iPSCs) and share similar characteristics with ESCs[10]. Notably, there has been no pluripotent cell population isolated from the lung.

3.1.3 Multipotent Stem Cells

Multipotent stem cells are found in most tissues and are able to differentiate into cells from a single germ layer. Mesenchymal stem cells (MSCs) are the most recognized multipotent stem cell. They can be derived from a variety of tissue including bone marrow, adipose tissue, bone, umbilical cord blood, and peripheral blood. MSCs are adherent to cell culture dishes and are characterized by specific surface cell markers. These cells can differentiate into mesoderm-derived tissue such as adipose tissue, bone, cartilage, and muscle. Recently, MSCs were differentiated into neuronal tissue which is derived from the ectoderm. This is an example of transdifferentiation, when a cell from one germ layer (mesoderm) differentiates into neuronal tissue (ectoderm). Tissue-resident MSCs have been isolated from the lung; however, no other multipotent cell has been isolated to date.

3.1.4 Oligopotent Stem Cells

Oligopotent cells are able to self-renew and form 2 or more lineages within a specific tissue; for example, the ocular surface of the pig, including the cornea, has been reported to contain oligopotent cells that generate individual colonies of corneal and conjunctival cells. Hematopoietic stem cells are a typical example of oligopotent stem cells, as they can differentiate into both myeloid and lymphoid lineages. In the lung, studies suggest that bronchoalveolar duct junction cells may give rise to bronchiolar epithelium and alveolar epithelium[14].

3.1.5 Unipotent Stem Cells

Unipotent stem cells can self-renew and differentiate into only one specific cell type and form a single lineage such as muscle stem cells, giving rise to mature muscle cells and not any other cells. In the lung, type Ⅱ pneumocytes of the alveoli give rise to type Ⅰ pneumocytes[15].

3.2 Stem Cell Classification Based on Origin

Stem cells can be grouped into 4 broad categories based on their origin: ESCs, adult stem cells, tissue-resident stem cells, and iPSCs. In general, ESCs and iPSCs are pluripotent, whereas adult stem cells are oligopotent or unipotent.

3.2.1 Embryonic Stem Cells

ESCs are pluripotent, derived from the ICM of the blastocyst, a stage of the preimplantation embryo, 5 – 6 days post-fertilization. These cells can differentiate into tissue of the 3 primary germ layers but can also be maintained in an undifferentiated state for a prolonged period in culture. The blastocyst has 2 layers of cells, the ICM, which will form the embryo, and the outer cell mass, called trophoblasts, that will form the placenta. Cells from the inner cell layer are separated from trophoblasts and transferred to a culture dish under very specific conditions to develop ESC lines. These factors maintain the stem cells in an undifferentiated state, capable of self-renewal. ESCs that have been cultured in an undifferentiated state with no genetic abnormalities are propagated as an ESC line. These cells could be frozen and thawed for further cultures and experimentation.

3.2.2 Adult Stem Cells

Adult stem cells are derived from adult tissue. Examples include MSCs as well as stem cells derived from placental tissue such as human amnion epithelial cells.

These cells have been shown to be anti-inflammatory and augment repair of animal models of injury. They have limited differentiation capacity although these cells have been differentiated into tissue from different germ cell layers *in vitro*.

Adult stem cells are of advantage since autologous cells do not raise issues of rejection or ethical controversies. Adult stem cells could be obtained from all tissues of the 3 germ layers as well as placenta. Several studies have demonstrated that transplantation of adult stem cells restores damaged organs *in vivo*, such as bone tissue repair and revascularization of the ischemic cardiac tissue via stem cell differentiation and generation of new specialized cells[13]. Other studies have shown that cultured adult stem cells secrete various molecular mediators with anti-apoptotic, immunomodulatory, angiogenic, and chemoattractant properties that promote repair.

3.2.3 Tissue-Resident Stem Cells

The ability of some tissues and organs in the adult to renew and repair following injury is critically dependent on tissue-resident stem cells that generate tissue-specific, terminally differentiated cells. Studies suggest that these cells originate during ontogenesis and remain in a quiescent state till local stimuli activate their proliferation, differentiation or migration.

Tissue-resident stem cells reside in a "stem cell niche". The stem cell niche is a microenvironment that controls the self-renewal and differentiation of these cells. There is a growing body of evidence that stem cell function is critically influenced by extrinsic signals from the microenvironment; therefore, the niche plays a crucial role in stem cell homeostasis and tissue repair. The majority of tissue-resident stem cells are dormant but are activated by specific signals during injury and repair. This dormancy of tissue-resident stem cells is not well understood but is likely influenced by the niche environment. This property is critical to maintaining a population of cells that do not perform other functions apart from generating tissue-specific cells during repair. The niche environment consists of various signals from extracellular matrix and soluble mediators that mediate cell signaling and gene expression, thereby regulating stem cell proliferation, migration, differentiation, or apoptosis. We still need to elucidate what the triggers are for stem cells to move from a self-renewal state and proliferation to differentiation and whether these signals are tissue specific.

Furthermore, the types of cell division that a stem cell undergoes determine the cells the type of cells generates. Symmetrical cell division by a stem cell results in identical daughter cells, which provides new cells to restitute damaged cells following injury. It is important to note that an uncontrolled increase in stem cell proliferation could lead to stem cell hyperplasia and/or carcinogenesis while a reduction in stem cells would impair organ repair; thus, balance in stem cell homeostasis is very important.

Asymmetric division occurs when a stem cell generates an identical daughter cell and a second differentiated daughter cell. This process allows for organ repair and regeneration while maintaining a population of stem cells.

3.2.4 Induced Pluripotent Stem Cells

iPSCs are produced from adult somatic cells that are genetically reprogrammed to an "ESC-like state".

Mouse iPSCs were reported for the first time by Takahashi and Yamanaka in 2006 by transducing mouse fibroblasts with 4 genes encoding the following transcription factors: octamer-binding transcription factor 3/4 (Oct-3/4), SRY-related high-mobility group box protein-2 (Sox2), the oncoprotein c-Myc, and Kruppel-like factor 4 (Klf4). A year later, in 2007, Yamanaka and colleagues described the generation of human iPSCs from adult human dermal fibroblasts with the same 4 factors: Oct-3/4, Sox2, Klf4, and c-Myc. They demonstrated that these cells were similar to human ESCs in terms of morphology, proliferation, surface antigens, gene expression, epigenetic status of pluripotent cell-specific genes, and telomerase activity, and they could differentiate into cell types of the 3 germ layers *in vitro*[16]. iPSCs are currently useful tools for drug development, modeling of diseases, and regenerative medicine, but although these cells express identical characteristics of pluripotent stem cells, it is not yet known if iPSCs and ESCs would significantly differ in clinical practice.

Retroviral vectors, used to introduce the reprogramming factors into adult cells, and oncogenes like c-Myc limit the use of iPSCs in a clinical study since the vectors used to introduce transcription factors to adult cells can cause cancer. Researchers are currently investigating new methods to generate safe iPSCs without genomic manipulation. New techniques have been described, using several types of mouse and human adult somatic cells. To avoid the use of oncoproteins c-Myc and Klf4, they have used one factor (Oct-3/4 or Klf4) or they have substituted them with combinations of other factors, including the use of non-retroviral vector approaches, such as chemical compounds, plasmids, adenovirus, and transposons.

Despite the safety issues, this innovative discovery has created a powerful tool to reprogram somatic adult cells by "sending them back" to earlier undifferentiated stages and generating iPSCs, thereby creating an identical match to the cell donor and thus avoiding issues of rejection.

4 Stem Cells and Their Niches

Stem cells, as functional units of growth and regeneration in many tissues, are important for maintaining proper tissue function. Thus, these cells should be protected as much as possible from damage or loss, while at the same time maintaining sufficient communication with their surroundings to ensure appropriate responsiveness to physiological cues for cell replacement and repair.

In many tissues, this balance between protection and interaction appears to be accomplished by maintaining stem cells in a specialized microenvironment, or niche, which provides spatial and temporal cues to support and coordinate stem cell activities. Stem cell niches have been identified and characterized in many tissues, including the germline, bone marrow, digestive and respiratory systems, skeletal muscle, skin, hair follicle, mammary gland, and the central and peripheral nervous systems.

Extensive studies in many different laboratories have begun to elucidate the critical components of many stem cell niches, which include specific mesenchymal, vascular, neuronal, glial, and inflammatory cell types, diffusible and cell surface-associated signaling molecules, and physical parameters such as matrix rigidity, shear stress, oxygen tension, and temperature. In particular, cell-cell interactions within the niche provide structural support, regulate adhesive interactions, and produce soluble signals that can control stem cell function[17].

Stem cell interactions with the extracellular matrix (ECM) provide retention cues, as well as mechanical signals, based in part on substrate rigidity, which allow stem cells to respond to external physical forces. In addition, the ECM can sequester or concentrate growth factors, chemokines, and other stem cell regulatory molecules by binding both locally and systemically produced factors within the niche. The close association of many stem cell types with the vasculature and nervous system allows for modulation of stem cell responses by metabolic cues and circadian rhythms, and provides a conduit through which inflammatory and immune cells, as well as humoral factors, can be delivered to the niche[18].

Finally, temperature, shear forces, and chemical signals provided by the niche also influence stem cell behavior in response to the external environment. Importantly, while the specific components that constitute a particular stem cell niche may vary in different tissues and under distinct physiological contexts, in all cases the signals provided by these cellular and acellular components appear to be integrated by stem cells to inform their fate decisions, including choices between quiescence or proliferation, self-renewal or differentiation, migration or retention, and cell death or survival.

Because the niche, by definition, impacts stem cell function extrinsically, it may be argued that this anatomical structure represents an even more "druggable" target for regenerative medicine than the stem cell itself.

5 Application of Stem Cells

It is possible to expand and induce stem cells *in vitro* for targeted differentiation, develop a single-type stem cell expansion method that meets clinical standards, and study the growth, migration, and differentiation of stem cells after transplantation into the human body, and even the reconstruction of tissue and organ structure and function.

In clinical applications, scientists have successfully differentiated human embryonic stem cells into hepatocytes, endothelial cells, cardiomyocytes, pancreatic cells, hematopoietic cells and neurons *in vitro*.

In terms of tissue stem cells, scientists can successfully isolate and culture stem cells from skin, bone, bone marrow, and adipose tissues, and try to use these cells for disease treatment. Using stem cells to construct various tissues and organs and using them as the source of transplantation will become the application direction of stem cells[19].

The growing maturity of stem cell, tissue and organ transplantation technology has played a huge role in promoting the development of regenerative medicine. In recent years, stem cells have been used in the following areas.

5.1 Stem Cells Used in Regenerative Medicine

Stem cells have great potential to become one of the most important aspects of medicine. In addition to the fact that they play a large role in developing restorative medicine, their study reveals much information about the complex events that happen during human development.

The difference between a stem cell and a differentiated cell is reflected in the cells' DNA. In the former cell, DNA is arranged loosely with working genes. When signals enter the cell and the differentiation process begins, genes that are no longer needed are shut down, but genes required for the specialized function will remain active.

Many serious medical conditions, such as birth defects or cancer, are caused by improper differentiation or cell division. Currently, several stem cell therapies are possible, among which are treatments for spinal cord injury, heart failure, retinal and macular degeneration, tendon ruptures, and diabetes type 1. Stem cell research can further help in better understanding stem cell physiology. This may result in finding new ways of treating currently incurable diseases[20].

5.1.1 To Treat Blood-Related Diseases

The process of autologous bone marrow transplantation is as follows: stem cells are collected from the patient's bone marrow or blood. Blood or bone marrow is processed in the laboratory to purify and concentrate the stem cells. Next, blood or bone marrow is frozen to preserve it. High dose chemotherapy and/or radiation therapy is given to the patient. The final thawed stem cells are reinfused into the patient.

5.1.2 Stem Cells as an Alternative for Arthroplasty

One of the biggest fears of professional sportsmen is getting an injury, which most often signifies the end of their professional career. This applies especially to tendon injuries, which, due to current treatment options focusing either on conservative or surgical treatment, often do not provide acceptable outcomes. Problems with the tendons start with their regeneration capabilities. Instead of functionally regenerating after an injury, tendons merely heal by forming scar tissues that lack the functionality of healthy tissues. Factors that may cause this failed healing response include hypervascularization, deposition of calcific materials, pain, or swelling.

In addition to problems with tendons, there is a high probability of acquiring a pathological condition of joints called osteoarthritis (OA). OA is common due to the avascular nature of articular cartilage and its low regenerative capabilities. Although arthroplasty is currently a common procedure in treating OA, it is not ideal for younger patients because they can outlive the implant and will require several surgical procedures in the future. These are situations where stem cell therapy can help by stopping the onset of OA. However, these procedures are not well developed, and the long-term maintenance of hyaline cartilage requires further research. Osteonecrosis of the femoral hip (ONFH) is a refractory disease associated with the collapse of the femoral head and risk of hip arthroplasty in younger populations. Although total hip arthroplasty (THA) is clinically successful, it is not ideal for young patients, mostly due to the limited lifetime of the prosthesis. An increasing number of clinical studies have evaluated the therapeutic effect of stem cells on ONFH. Most of the authors demonstrated positive outcomes, with reduced pain, improved function, or avoidance of THA.

5.1.3 Rejuvenation by Cell Programming

Ageing is a reversible epigenetic process. The first cell rejuvenation study was published in 2011. Cells from aged individuals have different transcriptional signatures, high levels of oxidative stress, dysfunctional mitochondria, and shorter telomeres than in young cells. There is a hypothesis that when human or mouse adult somatic cells are reprogrammed to iPSCs, their epigenetic age is virtually reset to zero. This was based on an epigenetic model, which explains that at the time of fertilization, all marks of parenteral ageing are erased from the zygote's genome and its ageing clock is reset to zero.

In recent study, scientists used Oct-4, Sox2, Klf4, and c-Myc genes (OSKM genes) and affected pancreas and skeletal muscle cells, which have poor regenerative capacity. Their procedure revealed that these genes can also be used for effective regenerative treatment. The main challenge of their method was the need to employ an approach that does not use transgenic animals and does not require an indefinitely long application. The first clinical approach would be preventive, focused on stopping or slowing the ageing rate.

Later, progressive rejuvenation of old individuals can be attempted. In the future, this method may raise some ethical issues, such as overpopulation, leading to lower availability of food and energy[21].

For now, it is important to learn how to implement cell reprogramming technology in non-transgenic elder animals and humans to erase marks of ageing without removing the epigenetic marks of cell identity[11].

5.1.4 Other Potential Medical Uses of Stem Cells

In brief, stem cells created from patient cells have the following characteristics when used in disease:

(1) Repair the disease-related genetic defects, and then healthy cells are transplanted to patients.

(2) Obtain new cells and tissues to repair damaged organs/tissues.

(3) Without immune rejection or low immune rejection.

5.2 Stem Cells for Drug Screening and Development in Laboratory

In the laboratory, through stem cell culture technology, new cells or tissues are obtained for various experiments.

Test new drugs for safety and effectiveness. Stem cells, as the target for pharmacological testing, can be used in new drug tests. Each experiment on living tissue can be performed safely on specific differentiated cells from pluripotent cells. If any undesirable effect appears, drug formulas can be changed until they reach a sufficient level of effectiveness. The drugs can enter the pharmacological market without harming any live testers. However, to test the drugs properly, the conditions must be equal when comparing the effects of two drugs. To achieve this goal, researchers need to gain full control of the differentiation process to generate pure populations of differentiated cells.

5.3 A Multidisciplinary Field

Regenerative medicine combines knowledge from "regenerative biology" with existing and new technologies (Fig. 1-3).

Regenerative biology
- Disease knowledge
- Tissue growth, damage and repair
- Cellular differentiation
- Molecular Biology
- Stem cell biology
- Developmental biology

Engineering and technology
- Biochemistry
- Biomaterials
- Bioreactors
- Biomechanics
- Imaging
- Implantation technology
- Nanotechnology
- Cell/protein engineering
- Genomics/proteomics

Fig. 1 -3 Stem cell in multidisciplinary field

6 Challenges and Requirements

Pioneering scientific and medical advances always have to be carefully policed in order to make sure they are both ethical and safe. Although stem cells appear to be an ideal solution for medicine, there are still many obstacles that need to be overcome in the future. One of the challenges is ethical concern.

The most common pluripotent stem cells are ESCs. Therapies concerning their use at the beginning were, and still are, the source of ethical conflicts. The reason behind it started when, in 1998, scientists discovered the possibility of removing ESCs from human embryos. Stem cell therapy appeared to be very effective in treating many, even previously incurable, diseases. The problem was that when scientists isolated ESCs in the lab, the embryo, which had potential for becoming a human, was destroyed. Because of this, scientists, seeing a large potential in this treatment method, focused their efforts on making it possible to isolate stem cells without endangering their source—the embryo. For now, the only ethically acceptable operation is an injection of hESCs into mouse embryos in the case of pluripotency evaluation.

The efficiency of stem cell-directed differentiation must be improved to make stem cells more reliable and trustworthy for a regular patient. The scale of the procedure is another challenge. Future stem cell therapies may be a significant obstacle. Transplanting new, fully functional organs made by stem cell therapy would require the creation of millions of working and biologically accurate cooperating cells. Bringing such complicated procedures into general, widespread regenerative medicine will require interdisciplinary and international collaboration.

The identification and proper isolation of stem cells from a patient's tissues is another challenge. Immunological rejection is a major barrier to successful stem cell transplantation. With certain types of stem cells and procedures, the immune system may recognize transplanted cells as foreign bodies, triggering an immune reaction resulting in transplant or cell rejection. One of the ideas that can make stem cells a "failsafe" is about implementing a self-destruct option if they become dangerous. Further development and versatility of stem cells may cause reduction of treatment costs for people suffering from currently incurable diseases[22].

Other challenges related to stem cells include:

(1) Safety of stem cell therapies needs to be carefully assessed.

(2) Even though autologous stem cells should not cause an immune response, there are other risks.

A. Contamination with pathogens.

B. Changes in the cells themselves, e. g. proliferation.

C. Introduction back to a different environment within the body than the original tissue may have unexpected consequences.

(3) Gene therapies may have unexpected side effects.

A. Random integration of viral genome.

B. CRISPR/Cas will be more specific.

(4) Problem with ESCs and iPSCs: risk of teratoma formation.

(5) Teratoma: usually benign germ cell tumor.

A. Teratoma formation a way to test pluripotency of stem cells.

B. Adult stem cells don't form teratomas since "only" multipotent.

Although these challenges facing stem cell science can be overwhelming, the field is making great advances each day. Stem cell therapy is already available for treating several diseases and conditions.

Supplement

List of Abbreviations	
CLP	common lymphoid progenitor
CMP	common myeloid progenitor
ECM	extracellular matrix
EpSCs	epithelial stem cells
ESCs	embryonic stem cells
hESCs	human embryonic stem cells
HSCs	hematopoietic stem cells

(To be continued)

List of Abbreviations	
iPSCs	induced pluripotent stem cells
MULTI	multipotent
NSCs	neural stem cells
OA	osteoarthritis
OLIGO	oligopotent
ONFH	osteonecrosis of the femoral hip
PLURI	pluripotent
SSCs	somatic stem cells
THA	total hip arthroplasty
TOTI	totipotent

Key Words List	
成体干细胞	adult stem cell
单能干细胞	unipotent stem cell
多能干细胞	pluripotent stem cell
固有的，同源的	autologous
寡能干细胞	oligopotent stem cell
胚胎干细胞	embryonic stem cell
全能干细胞	totipotent stem cell
细胞外基质	extracellular matrix
专能干细胞	multipotent stem cell
组织常驻干细胞	tissue-resident stem cell

References

[1] TILL J E, MCCULLOCH E A. A direct measurement of the radiation sensitivity of normal mouse bone marrow cells. 1961 [J]. Radiation research, 2011, 175 (2): 145-149.

[2] WILMUT I, SCHNIEKE A E, MCWHIR J, et al. Viable offspring derived from fetal and adult mammalian cells [J]. Nature, 1997, 385: 810-813.

[3] THOMSON J A, ITSKOVITZ-ELDOR J, SHAPIRO S S, et al. Embryonic stem cell lines derived from human blastocysts [J]. Science, 1998, 282: 1145-1147.

[4] TAKAHASHI K, TANABE K, OHNUKI M, et al. Induction of pluripotent stem cells from adult human fibroblasts by defined factors [J]. Cell, 2007, 131: 861-872.

[5] CODEGA P, SILVA-VARGAS V, PAUL A, et al. Prospective identification and purification of quiescent adult neural stem cells from their in vivo niche [J]. Neuron, 2014, 82: 545 – 559.

[6] SOUSA B R, PARREIRA R C, FONSECA E A, et al. Human adult stem cells from diverse origins: an overview from multiparametric immunophenotyping to clinical applications [J]. Cytometry. Part A : the journal of the International Society for Analytical Cytology, 2014, 85: 43 – 77.

[7] BOND A M, MING G L, SONG H. Adult mammalian neural stem cells and neurogenesis: five decades later [J]. Cell stem cell, 2015, 17: 385 – 395.

[8] LIU Z, CAI Y, WANG Y, et al. Cloning of macaque monkeys by somatic cell nuclear transfer [J]. Cell, 2018, 172: 881 – 887. e7.

[9] LIU G, DAVID B T, TRAWCZYNSKI M, et al. Advances in pluripotent stem cells: history, mechanisms, technologies, and applications [J]. Stem cell reviews and reports, 2020, 16: 3 – 32.

[10] ZAKRZEWSKI W, DOBRZYNŃSKI M, SZYMONOWICZ M, et al. Stem cells: past, present, and future [J]. Stem cell research & therapy, 2019, 10: 68.

[11] TAKAHASHI K, YAMANAKA S. Induced pluripotent stem cells in medicine and biology [J]. Development, 2013, 140: 2457 – 2461.

[12] DAVIS R L, WEINTRAUB H, LASSAR A B. Expression of a single transfected cDNA converts fibroblasts to myoblasts [J]. Cell, 1987, 51: 987 – 1000.

[13] OH I H, KIM D W. Three-dimensional approach to stem cell therapy [J]. Journal of Korean medical science, 2002, 17: 151 – 160.

[14] LIN H T, OTSU M, NAKAUCHI H. Stem cell therapy: an exercise in patience and prudence [J]. Philosophical Transactions of the Royal Society of London. Series B, biological sciences, 2013, 368: 20110334.

[15] KOLIOS G, MOODLEY Y. Introduction to stem cells and regenerative medicine [J]. Respiration, 2013, 85: 3 – 10.

[16] TAKAHASHI K, TANABE K, OHNUKI M, et al. Induction of pluripotent stem cells from adult human fibroblasts by defined factors [J]. Cell, 2007, 131: 861 – 872.

[17] WANG X. Stem cells in tissues, organoids, and cancers [J]. Cellular and molecular life sciences, 2019, 76: 4043 – 4070.

[18] WAGERS A J. The stem cell niche in regenerative medicine [J]. Cell stem cell, 2012, 10: 362 – 369.

[19] DULAK J, SZADE K, SZADE A, et al. Adult stem cells: hopes and hypes of regenerative medicine [J]. Acta biochimica polonica, 2015, 62: 329 – 337.

[20] BIEHL J K, RUSSELL B. Introduction to stem cell therapy [J]. Journal of cardiovascular nursing, 2009, 24: 98 – 105.

[21] VOLAREVIC V, LJUJIC B, STOJKOVIC P, et al. Human stem cell research and regenerative medicine-present and future [J]. British medical bulletin, 2011, 99: 155 -168.

[22] MAJKA M, KUCIA M, RATAJCZAK MZ. Stem cell biology: a never ending quest for understanding [J]. Acta biochimica polonica, 2005, 52: 353 -358.

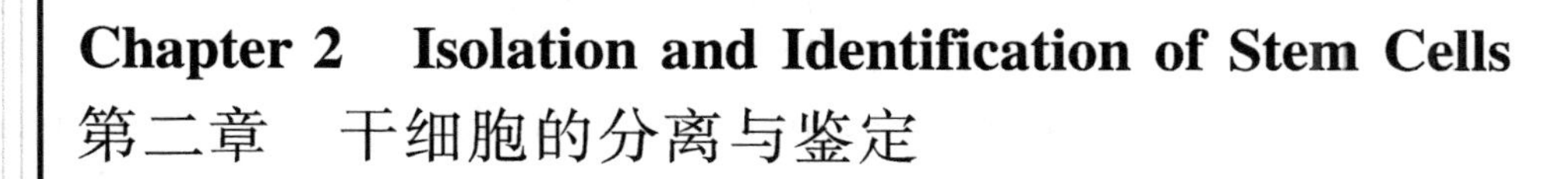

Chapter 2 Isolation and Identification of Stem Cells
第二章　干细胞的分离与鉴定

［**中文导读**］

干细胞是具有自我更新及多向分化潜能的细胞，是组织器官再生的种子细胞，是实践再生医学的最重要的先决条件。能够迅速、精确地分离和鉴定不同种类的干细胞，是研究干细胞的前提和基础。

胚胎干细胞特指哺乳类动物着床前囊胚内细胞团在体外特定条件下培养和扩增所获得的永生性细胞。它们能在体外培养中无限自我更新并具有分化为各种类型的细胞和在体内形成畸胎瘤的能力。胚胎干细胞的分化，可以作为体外模型研究人类胚胎发育过程；利用胚胎干细胞可以分化成特定细胞和组织的特性，可建立人类疾病模型，用于疾病的发病机制研究、药物筛选和研发等；利用胚胎干细胞分化出来的供体细胞，还可以针对目前难治的疾病开展细胞移植治疗。

间充质干细胞最早是从人类骨髓中分离出来的，骨髓间充质干细胞被用做中胚层分化的模型。为了改进间充质干细胞的临床应用，需要进一步阐明间充质干细胞的组织修复和免疫抑制的作用机制；同时，需要对临床研究中的间充质干细胞分离和培养扩增方法进行标准化。

诱导多能干细胞是利用反转录病毒基因表达载体，将已知在胚胎干细胞中高表达的 4 种转录因子（Oct4、Sox2、Klf4 和 c-Myc）导入胎鼠或者成年小鼠的皮肤成纤维体细胞中，在体外直接将这些分化的细胞诱导成为类似胚胎干细胞的多能干细胞。诱导多功能干细胞具有与胚胎干细胞相似的无限的自我更新能力、分化的多能性和在体内形成畸胎瘤等特点。由于其来源于成体细胞，不存在破坏胚胎获取干细胞而带来的伦理问题。

成体干细胞具有自我复制能力，并能产生不同种类的具有特定表型和功能的成熟细胞，能够维持机体功能的稳定，发挥生理性的细胞更新和修复组织损伤作用。一般将培养扩增的成体干细胞作为自体或异体干细胞移植的细胞来源。成体干细胞一般不存在成瘤和伦理学问题，并且同样具有很强的多潜能性，自体移植不存在免疫排斥。本章详细介绍成体干细胞中的成体神经干细胞。

1 Embryonic Stem Cells

One of the great advantages of embryonic stem cells (ESCs) over other cell types is their accessibility of genetic manipulation. They can easily undergo genetic modifications while remaining pluripotent, and can be selectively propagated, allowing the clonal expansion of genetically altered cells in culture. Since the first isolation of ESCs in mice, many effective techniques have been developed for gene delivery and manipulation of ESCs. These include transfection, electroporation, and infection protocols, as well as different approaches for inserting, deleting, or changing the expression of genes. These methods are proved to be extremely useful in mouse ESCs, for monitoring and directing differentiation, discovering unknown genes, and studying their function, and are now being extensively implemented in hESCs. This section describes the different approaches and methodologies that have been applied for the genetic manipulation of hESCs and their applications. Detailed protocols for generating clones of genetically modified hESCs by transfection, electroporation, and infection will be described, with special emphasis on the important technical details that are required for this purpose. All protocols are equally effective in human-induced pluripotent stem cells (iPSCs).[1]

1.1 Genetic Modification Approaches and Potential Applications

There are basically two types of strategies that can be applied for inducing permanent changes in the DNA of hESCs. One approach depends on random integration of foreign DNA sequences into the genome while the other approach relies on targeted mutagenesis.

1.1.1 Random Integration of Foreign Sequences into the Genome

Random integration of foreign sequences into the genome is typically applied for overexpression of genes, or for the downregulation of endogenous genes in *trans* (knock-down). Overexpression is usually useful for constitutive or facultative expression of either cellular or foreign genes. It may also be applied for the introduction of reporter or selection genes, under the regulation of tissue-specific promoters. These procedures allow of labelling and tracking specific cell lineages following induced differentiation of hESCs in culture. Moreover, they can be employed for the isolation of pure populations of specific cell types, by the use of selectable markers.

The marker gene may either be a selectable reporter, such as green fluorescent protein (GFP), resulting in the production of green glowing cells which can be selected for by fluorescent activated cell sorter (FACS), or a drug resistance gene[2]. The ability to isolate pure populations of specific cell types and eliminate undifferentiated cells prior to transplantation has great importance in cell-based therapy; this is because transplantation of undifferentiated cells may lead to teratoma formation. Overexpression experiments may also be employed for directing the cell fate of differentiating ESCs in culture. This can be achieved by introducing master genes that play a dominant role in cell commitment, forcing the cells to differentiate into specific lineages that otherwise are rarely obtained among many other cell types in culture. Random integration of promoter-driven transgenes may also be employed for the generation of cell-based delivery systems by producing therapeutic agents at the site of damaged tissue. The use of embryonic stem-derived cells as therapeutic vectors has been previously shown to be feasible in mice, where grafting of embryonic stem-derived insulin secreting cells normalized glycemia in streptozotocin-induced diabetic mice[3].

Apart from tagging, selecting, and directing the differentiation of specific cell types, it is possible to inactivate endogenous genes to study their function. This can be achieved by downregulating the activity of particular genes in *trans* by overexpressing specific short hairpin RNA (shRNA) molecules. shRNAs are short sequences of RNA that by forming hairpins silence target gene expression via RNA interference (RNAi) pathway. They are processed into small interfering RNAs (siRNAs) by the enzyme Dicer, and then paired with the target mRNA as they are incorporated into an RNA-induced silencing complex (RISC), leading to the degradation of the target mRNA. The great advantage of this system is that it provides a specific, long-lasting, gene silencing effect. This is why it is being considered as one of the most applicable tools for gene silencing in living organisms. Furthermore, since shRNAs operate in *trans* and are not involved in the modification of the targeted gene, it is relatively simple to apply and particularly efficient in achieving transient or conditional gene silencing effects. Expression of shRNA in hESCs is typically accomplished by transfection or through viral infection. Applications of this loss-of-function approach are now widely used not only to study developmental roles of specific genes in human, but also for their utility in modulating hESC differentiation *in vitro*[4].

An additional use for the random integration approach can be the search of unknown genes whose pattern of expression suggests that they might have developmental importance. The identification of such genes is performed by the gene trap method, which is based on the random disruption of endogenous genes.

As opposed to targeted mutagenesis (see below), it involves the random insertion of a reporter gene that lacks essential regulatory elements into the genome. Because the expression of the reporter gene is conditioned by the presence of an active endogenous regulatory element, it may serve to identify only transcribed sequences.

Using this method, a large-scale gene disruption assay is possible, allowing the discovery of new genes and the creation of wide variety of mutations[5].

1.1.2 Targeted Mutagenesis

Targeted mutagenesis, or site-directed mutagenesis, is a procedure which involves the replacement of a specific sequence in the genome by a mutated copy through homologous recombination with a targeting vector. The targeting vector that contains the desired mutation and a selectable marker, flanked by sequences that are interchangeable with the genomic target, pairs with the wild-type chromosomal sequence and replaces it through homologous recombination. Targeted mutagenesis is most widely used technique for inactivating genes in ESCs. By targeting both alleles, using distinct selection markers, it is possible to create "loss-of-function" or so-called knockout phenotypes in ESCs that can be used for functional studies of specific genes. This technology has been well practiced in mice for gene function studies, in which genetically altered cells are introduced into wild-type embryos, resulting in the creation of germ-line transmitting chimeras. The genetically manipulated animals can be further mutated to generate animals that are homozygous for the desired mutation. The creation of hESCs with a null genotype for specific genes may have great importance for modeling human diseases, and for the study of crucial developmental genes that in their absence are embryonic lethal[6]. Thus, these cells should be valuable for basic research studies, but more importantly for exploration of new gene therapy-based treatments and drug discovery.

A very similar approach that relies on targeted mutagenesis involves the insertion of foreign sequences into the genome at desired loci. This strategy, termed knock-in, is commonly used to study the regulatory function of specific elements, for example, by positioning a reporter gene under the regulation of a native gene[7].

Therefore, it can be applied to follow the expression of a target gene in situ during ESCs differentiation and monitoring the expression of the endogenous genes, enabling to identify hESCs differentiated cell derivatives. It should be emphasized that both gene targeting approaches, knock-out and knock-in, depend on homologous recombination events. However, the efficiencies of homologous recombination are extremely low, limiting the routine use of these techniques in hESC manipulation until recently. Yet, as double strand breaks dramatically improve the rate of homologous recombination, it was hypothesized that by targeting double strand DNA breaks to specific sites in the genome one may significantly improve the efficiencies of targeted mutagenesis.

Indeed, due to the recent advancements in the field of artificially engineered nucleases, it has been possible to insert, replace, or remove specific DNA sequences from the genome of hESCs/iPSCs in a fairly uncomplicated procedure. This technology, termed genome editing, depends on the direction of unspecific DNA nucleases to desired sites in the genome, where they induce double strand DNA breaks and by that significantly enhance the rate of homologous recombination. There are by now three different types of engineered nucleases that can be applied for this purpose: zincfinger nucleases (ZFNs), transcription activator-like effector nucleases (TALENs) and RNA-guided engineered nucleases (RGENs). All results in the elevation of gene targeting events through homologous recombination by at least 2 – 3 orders of magnitude relative to the conventional method by transient expression[8].

1.2 Methods for Genetic Manipulation

Several gene transfer techniques are now available for manipulating gene expression in hESCs, including chemical-based (transfection), physical-based (electroporation), and viral-mediated (infection) techniques. No single transfection method will work for all hESC lines, and even within a lab, the method of choice may vary.

1.2.1 Transfection

Transfection is probably the most commonly used method for introducing transgenes into hESCs. It is straightforward, relatively easy to calibrate, provides a sufficient number of cells for clonal expansion, which can be performed on adherent cell cultures, and allows the insertions of constructs of virtually unlimited size. This system is based on the use of carrier molecules that bind to foreign nucleic acids and introduce them into the cells through the plasma membrane. In general, the uptake of exogenous nucleic acids by the cell is thought to occur through endocytosis, or in the case of lipid based reagents, through fusion of lipid vesicles to the plasma membrane. There are many factors that may influence transfection efficiency: phase of cell growth, number of passages, size and source of the transgene, vector type and size, and selection system. However, the most important factor is the transfection method. The first study to describe stable transfection in hESCs was based on the use of a commercially available reagent, ExGen 500, which is a linear polyethylenimine (PEI) molecule that has a high cationic charge density[4]. The unique property of this molecule is due to its ability to act as a "proton sponge", which buffers the endosomal pH, leading to endosome rupture and DNA release. This method routinely produces transient transfection rates of approx 10% – 20% and stable transfection efficiencies of $1:10^{-5}$ to 10^{-6}. Since then, other chemical-based transfection methods have been found to be equally effective.

Both reagents are based on the presence of a positively charged cationic lipid compound that forms small unilamellar liposomes and are useful in obtaining transient and stable transfections in hESCs as well. Usually, the cells are plated to 50% – 70% confluence at the time of transfection. The plasmid DNA and lipid reagent are mixed in a tube, and only then administered to the cells as a DNA-lipid complex.

1.2.2 Electroporation

Electroporation is a method that employs the administration of short electrical impulses that create transient pores in the cell membrane, allowing foreign DNA to enter into the cells. Although it is efficient and most popular in mouse ESCs, this procedure gave poor results in hESCs, both in transient and stable transfection experiments. This is most probably due to the low survival rates of hESCs after the voltage shock. This was performed by carrying out the procedure on cell clumps rather than on single cell suspension. In addition, electroporation was performed in standard cell culture media, which is a protein-rich solution, instead of PBS and altering the parameters of the protocol used in mouse ESCs. Using this method, 3% –40% homologous recombination events among resistant clones were reported, subject to vector properties[4]. A substantial number of hESC clones obtained by homologous recombination have been created thus far using different constructs, demonstrating the feasibility of this technique for site-directed mutagenesis in hESCs.

1.2.3 Infection

Unlike in all nonviral-mediated methods (transfection and electroporation), gene manipulation by viral infection can produce a very high percentage of modified cells. To date, genetic manipulation of hESCs by viral infection has been reported by several groups using adeno as well as baculovirus and lentiviral vectors (LVVs). Infection studies with RNA and DNA viruses have demonstrated that these viral vectors have two distinct advantages over other systems: high efficiency of DNA transfer and single-copy integrations. However, integration occurs randomly and cannot be targeted to a specific site in the genome[9]. Yet, because of its high efficiency, this method could prove useful for bypassing the need for selection and time-consuming clonal expansion, as well as for experiments that aim for random insertion mutagenesis or gene trap.

Lentiviral-based vectors offer an attractive system for efficient gene delivery into hESCs. LVVs can transduce both dividing and nondividing cells and were shown to drive gene expression efficiently in various types of "stem" cells. Gene delivery into hESCs by vectors derived from lentiviruses has the following advantages:

(1) LVVs efficiently transduce hESCs.

(2) They integrate into the host cell genome, thus promoting stable transgene expression.

(3) Transgene expression is not significantly silenced in undifferentiated hESCs as well as following differentiation.

(4) Transduced hESCs retain their self-renewal and pluripotent potential.

To improve vector biosafety and performance, all pathogenic coding sequences were deleted, resulting in a replication-defective vector. In addition, the proteins necessary for the early steps of viral infection (entering into the host cell, reverse transcription, and integration) were provided in *trans* by two additional plasmids: a packaging plasmid expressing the gag, pol, and rev genes, and an envelope plasmid expressing a heterologous envelope glycoprotein of the vesicular stomatitis virus (VSV-G). Moreover, a large deletion was introduced to abolish the viral promoter/enhancer activity. These steps resulted in a vector that could only undergo one round of infection and integration, a process termed transduction. Furthermore, they minimized the risk of generation of wild-type HIV-1 by recombination.

Random chromosome integration of viral vectors poses the risk of insertional mutagenesis, oncogene activation, and cellular transformation. In addition, LVVs may not be suitable for transient transgene expression. Viral vectors derived from adenovirus and adeno-associated virus (AAV) have a much lower risk of insertional mutagenesis and have been tested in hESCs, but their transduction efficiencies were less satisfactory. The insect baculovi-rus autographa californica multiple nucleopolyhedrovirus (AcMNPV)-based vectors have also been introduced as a type of delivery vehicle for transgene expression in mammalian cells.

The virus can enter mammalian cells but does not replicate, and it is unable to recombine with preexisting viral genetic materials in mammalian cells. One significant advantage of using baculovirus AcMNPV as a gene delivery vector is the large cloning capacity to accommodate up to 30 kilobases (kb) of DNA insert, which can be used to deliver a large functional gene or multiple genes from a single vector[10].

1.2.4 Short-vs. Long-Term Expression

Gene transfer experiments can be subdivided into short-term (transient) and long-term (stable) expression systems. In transient expression, the foreign DNA is introduced into the cells and its expression is examined within 1 – 2 days. The advantage of this assay is its simplicity and rapidity. Furthermore, because the foreign DNA remains episomal, there are no problems associated with site of integration and the copy number of the transgene.

Yet, it does not allow conducting experiments over long periods. Moreover, transfection efficiency usually does not exceed 20%. For short-term induction, efficient transient expression can be achieved through the insertion of supercoiled plasmid DNA rather than the linear form. Transient expression in hESCs usually peaks roughly 48 h after transfection, and frequently results in high expression levels attributed to the high copy number of plasmid DNA molecules that occupy the cell.

During long-term assays, one isolates a clone of hESCs that has stably integrated the foreign DNA into its chromosomal genome. The major advantage of this method is the ability to isolate stable ESC lines that have been genetically modified and can be grown indefinitely in culture. In this type of experiment, it is important to linearize the vector, leading to greater integration and targeting efficiencies. When the target gene is non-selectable, one must introduce a positive selection marker as well under the regulation of a strong constitutive promoter. This can be performed either by co-transfecting the selectable marker on a separate vector, or as is frequently done, by fusing the selectable marker to the targeting vector. Selection should not be carried out immediately after transfection but at least 24 hours later, giving the cells time to recover, integrate the foreign DNA and express the resistance conferring gene.

2 Mesenchymal Stem Cells

MSCs, originally described in bone marrow, are the most widely recognized cell type for clinical use for several reasons: they are easy to obtain and cultivate in big numbers, they are capable of multilineage differentiation, they have the potential for autologous or allogeneic use, they have immunoregulatory features, and present few ethical dilemmas (with IRB approval and donor consent). MSCs can be isolated with high efficiency from various tissues, such as bone marrow, adipose tissue, umbilical cord, and deciduous teeth. The accepted definition of an MSC is that prescribed by the International Society for Cellular Therapy: an MSC must be plastic adherent, 95% or more of the cells of a colony must express CD105, CD73, and CD90, less than 2% of the cells may express CD45, CD434, CD14, CD11b, CD79α, CD19, or HLA class Ⅱ, and an MSC must be capable of forming cells of at least the osteogenic, chondrogenic, and adipogenic lineages[11].

MSCs have been shown to differentiate into mesodermal lineages (such as bone, fat, or cartilage), and have the ability to inhibit T-cell proliferation and activation, which together with their low immunogenicity suggests that they may have potential for allogeneic application. Unlike hematopoietic precursors, plastic-adherent MSCs can be expanded in culture without loss of differentiation potential, although their lifespan is limited to 15 – 50 population doublings. Moreover, MSCs lack a set of unique surface markers for identification, which minimally relies on the expression of several stromal markers and the ability to differentiate *in vitro* in response to particular stimuli.

There are currently more than 200 active clinical trials using MSCs. In order to maximize their therapeutic potential, a thorough understanding of MSC biology is required, necessitating high quality and characterized cell stocks. MSCs, together with HSCs, are the most frequently used cell type for cell-based therapeutics. As for other cell types intended for research and translational use, it is important to establish correctly typed cell lines from human tissue donations[12].

2.1 Mesenchymal Stem Cells Derived from Human Adipose Tissue

Similar to bone marrow-derived MSCs, adipose-derived stem cells (ASCs) are multipotent adult stem cells which, as their name suggests, can be isolated from adipose (fat) tissue. First identified in 2001, obtained via liposuction, ASCs were initially named processed lipoaspirate (PLA) cells. Morphologically similar to fibroblasts, upon differentiation, ASCs can give rise to adipocytes, osteocytes, and chondrocytes. This plasticity may offer potential for their use in regenerative medicine as large numbers of these daughter cells can be generated from an easily isolated stem cell source.

Furthermore, since aspirated fat is in plentiful supply from many plastic surgery procedures, such as liposuction and liposculpture, and the precursor cells can be purified by a variety of processing and enzymatic techniques to obtain the ASC-rich stromal vascular fraction (SVF), it is relatively straightforward to obtain ASCs. The availability of stem cells in fat is much greater than in bone marrow. MSCs comprise about 2% of nucleated cells in lipoaspirate compared with only 0.001% – 0.004% in bone marrow, the number of stem cells per milliliter of lipoaspirate is about eightfold higher than in bone marrow, and the volume of lipoaspirate typically obtained under local anesthesia is at least fivefold greater than is possible for bone marrow. The relative ease with which large numbers of ASCs can be obtained for tissue engineering applications is an important advantage.

Tissue-derived MSCs have many practical advantages for cartilage tissue engineering compared with other stem cell types. Use of MSCs avoids ethical concerns over embryo harvesting and the safety issues relating to tumor formation in recipient patients by embryonic and induced pluripotent stem cells. There is increasing evidence of an additional benefit of immune-privilege, in that allogeneic MSCs fail to activate host immune responses that are typically responsible for rejection of implanted cells and organs. This feature reflects the role of MSCs in early tissue repair and remodel when inflammation control is required; it also overcomes the immune-rejection problems associated with ESCs. Further research is required to understand the mechanisms behind the immunomodulatory and immunosuppressive functions of MSCs; nevertheless, these properties could open the way for clinical application of "off-the-shelf" replacement organs and tissues produced using cells that are not patient-specific[13].

ASCs cells have been shown to differentiate along classical mesenchymal lineages towards cartilage, bone, muscle, and fat, suggesting that ASCs have some degree of multilineage plasticity across different germ layers. Specific signaling molecules and culture conditions are required to achieve stem cell differentiation, extracellular matrix synthesis, and maintenance of the differentiated cell phenotype. For chondrogenesis and cartilage production, growth factors in the transforming growth factor β superfamily (TGF-βs) play an especially important role.

A further similarity between MSCs and ASCs is their promising immunomodulatory capacity, which has been observed to be either cell-contact dependent or mediated by cytokines and trophic factors such as TNF-α, IFN-γ, IDO, PGE-2, and IL-17. *In vivo*, ASCs have been used for this purpose in spinal cord injury and neurodegenerative diseases, allergic and autoimmune diseases (for example, rheumatoid arthritis and inflammatory bowel disease), and in reducing graft versus host disease (GVHD). Clinical trials are bringing the use of injectable and implantable ASCs closer to becoming a reality for patients.

However, ASCs have proven problematic to identify in culture, and studies have been carried out to point to particular cellular markers, which may make them easier to recognize. Previous work has characterized ASCs based on their morphology, cell surface marker expression, and/or by assessing their ability to differentiate into specific lineages. Once isolated, ASCs have an even, round phenotype, in contrast to the irregular shape of endothelial cells. When cultured, they adhere to plastic and assume a fibroblast-like morphology within several days[14].

Rigotti's groundbreaking work showed positive characterization by flow cytometry of

ASCs with respect to antibodies against CD105, CD73, CD29, CD44, and CD90. ASCs correlated negatively with CD31, CD45, CD14, and CD34 expression. These findings were consolidated by the International Society for Cellular Therapy (ISCT). [15]

ASCs can be isolated from human fat tissue using a combined approach of washing, centrifugation, and filtration followed by cell sorting based on their expression of the markers CD90, CD73, CD105, and CD44 and lack of expression for the markers CD45 and CD31. This method could be used to isolate and purify ASCs for downstream applications such as for regenerative medicine and reconstruction or for immune suppression studies[16].

2.2 Mesenchymal Stem Cells from Human Bone Marrow

The recent explosion of interest in developing cell and gene therapies using adult stem/progenitor cells from human bone marrow can be partly attributed to the ease of isolation and expansion of cells from this source *in vitro*. In addition, the possibility of generating genetically manipulated bone marrow-derived stem cells to introduce specific genes of interest makes them attractive vehicles for gene therapy.

Human MSCs (hMSCs) are readily isolated from bone marrow by their adherence to tissue culture plastic and can be expanded through multiple passages in medium containing high concentrations of fetal bovine serum (FBS). However, the proliferation rates and other properties of the cells gradually change during expansion, and therefore, it is advisable to not expand hMSCs beyond four or five passages, the most prominent properties of MSCs are their ability to generate colonies after they are plated at a low density, but both the colonies and the cells within a colony are heterogeneous in morphology, rates of proliferation, and efficacy with which they differentiate. Besides, cultures of expanded cells are heterogeneous in their content of cells possessing an early progenitor phenotype. hMSCs are highly sensitive to plating density, and early progenitors are rapidly lost if the cultures are grown to confluence. Although the most recent definition of MSCs includes the expression of CD105, CD90, and CD73 surface antigens as potential biomarkers for MSCs, they alone are not sufficient to isolate cells directly from human bone marrow. Therefore, it is important to devise standardized assays for isolating and characterizing MSCs[17].

For the primary isolation of bone marrow-derived MSCs, the critical steps include the isolation of mononucleated cells from a marrow aspirate by centrifugation on a density gradient followed by recovery and expansion of cells that adhere to tissue culture plastic in standard serum-containing medium (passage zero cells).

Passage zero cells are subsequently expanded by plating at a low density, which enhances the percentage of rapidly proliferating spindle-shaped cells. These cells would be replaced by large, flat, and thereby more mature hMSCs if the passage zero cells were plated at higher density or continually passaged for more than four to six times. Mature hMSCs will expand more slowly and have less multilineage differentiation potential, but still retain the ability to differentiate into mineralizing osteoblasts and secrete factors that enhance the growth of HSCs and perhaps other cells. The efficiency with which hMSCs form colonies still remains an important assay for the quality of cell preparations[17].

3 Induced Pluripotent Stem Cells

The isolation of murine ESCs in 1981 and the subsequent isolation of hESCs in 1998 has increased our understanding of normal embryonic development and opened the door for the development of stem cell derived treatments to debilitating diseases such as Type 1 diabetes and Parkinson's disease[18]. However, the isolation of ESCs requires the use of embryos, which is an ethically controversial approach. Therefore, new methods were needed to produce pluripotent stem cells without destroying embryos. This realization promoted the development of approaches involving the reprogramming of somatic cells. Although the field of cellular reprogramming is still in its infancy, the first successful reprogramming experiments took place nearly 60 years ago. The goal of these experiments, in which somatic cell nuclei from various amphibian species were transferred into enucleated eggs or zygotes, was not to reprogram cells but to determine if somatic cells possessed the full genetic complement of embryonic cells, as it had been previously hypothesized that as cells differentiated they would lose genetic material that was no longer needed. These first somatic cell nuclear transfer (SCNT) experiments were successful in demonstrating a somatic cell possesses a full genome and opened the door to the reprogramming of somatic cells to a pluripotent state.

Although it played a crucial role in our understanding of pluripotency and remains one of the most efficient methods available to reprogram somatic cells, SCNT is not an ideal strategy for producing pluripotent cells because it requires the use of an unfertilized egg. The isolation and manipulation of unfertilized eggs is commonly practiced in the setting of *in vitro* fertilization, but the scarcity of eggs in the research setting prohibits large scale use of SCNT. Thus, new technology is needed for the ethical and efficient generation of pluripotent stem cells.

In 2006, the Yamanaka group published a pioneering study in which they described reprogramming mouse cells with a defined cocktail of four transcription factors[19]. Soon after, both the Yamanaka and Thomson groups published studies describing the generation of human induced pluripotent stem cells (hiPSCs) using a four transcription factors cocktail consisting of Oct4 (O), Sox2 (S), Klf4 (K), and c-Myc (M) or Oct4, Sox2, Nanog (N), and Lin28 (L), respectively[20]. The iPSCs generated by these methods have been demonstrated to be very similar to ESCs (in morphology, pluripotency, mRNA profile, and protein expression) and many believe iPSCs will one day replace ESC as the pluripotent cell of choice for the development of clinical therapies.

Unlike ESCs, iPSCs generated using the retroviral transduction protocols described by the Yamanaka or Thomson groups contain transgenes that are known oncogenes[18, 21].

These transgenes have been introduced via retroviral transduction and contain multiple viral integrations, thus introducing the possibility for dysregulation of tumor suppressors or proto-oncogenes. The integration of these oncogenic transgenes makes traditional iPSCs poor candidates for clinical use. For this reason, numerous groups have developed new methods to produce integration-free iPSCs. The reprogramming process is time consuming, with traditional methods requiring 21 – 28 days.

Further, the whole process is highly inefficient, as only 0.002% of cells are successfully reprogrammed. This section describes traditional reprogramming methods and those designed to decrease the introduction of exogenous genetic material and increase reprogramming efficiency[22].

3.1 Integrating Vectors

3.1.1 Retroviral Transduction

The first experiment to describe the generation of pluripotent stem cells without the need to sacrifice eggs or embryos employed retroviral transduction to deliver reprogramming factors. Surprisingly only a "four transcription factor cocktail" is necessary to reprogram somatic cells. The efficiency of four factors reprogramming is low and has been shown to be improved with the inclusion of other factors such as Glis1, SV40LT, RARα, and LRH-1.

Multiple retroviral vectors were originally needed to deliver the reprogramming cocktail. Now a single vector, employing picornaviral 2A sequences can deliver all of the reprogramming vectors using a single virus. This allows fewer integration and therefore reduces the likelihood of disturbing cellular homeostasis.

As described above, the initial reprogramming experiments used retroviral transduction to deliver reprogramming factors. Random integration of the reprogramming cassette can result in dysregulation of proto-oncogenes or aberrant expression of mutated protein sequences. Additionally, integrated transgenes may not be completely silenced, resulting in difficulty differentiating reprogrammed cells or an increased risk of developing neoplasia. These hurdles must be overcome before cells or tissues derived from reprogrammed somatic cells can be developed for clinical use. Therefore, an ideal integrative system would allow for the removal of transgenes post reprogramming.

3.1.2 Excisable Reprogramming Cassettes

The first method used to excise integrated reprogramming cassettes take advantage of the Cre/LoxP system in the context of lentiviral transduction and integrative linear DNA transfection. Specifically, the integrated reprogramming cassette is flanked by loxP recombination sequences such that Cre expression in reprogrammed cells will excise the transgene cassette.

Although these methods will successfully remove the reprogramming factors, excision using Cre/loxP leaves a single loxP site behind potentially disrupting endogenous protein expression.

Recently, the Woltjen group has developed a system employing the piggyBac transposase. Like Cre/loxP, the piggyBac system effectively excises transgene sequences flanked by piggyBac recognition sequences. However, unlike Cre/loxP and other transposases, transposon excision utilizing piggyBac is clean because no exogenous DNA footprint is left behind. The piggyBac system utilizes piggyBac transposase to both introduce and remove a polycistronic reprogramming cassette. It is advantageous in that the reprogramming transgenes can be successfully removed without leaving a trace. However, reprogramming with piggyBac vectors requires an extra step to remove integrated transgenes, and extensive screening of iPSC colonies must be performed to ensure they are integration free. Even the most sensitive screening techniques may miss small single or oligomeric nucleic acid changes at excised integration sites. Therefore, the field has developed non-integrative methods to reprogram cells.

3.2 Non-DNA Reprogramming

3.2.1 Modified mRNA

One non-integrative, non-DNA method used to successfully reprogram somatic cells moves down the central dogma of molecular biology and employs modified mRNA to express four reprogramming factors, therefore bypassing the need for DNA and avoiding the dangers associated with DNA integration. This method uses *in vitro* transcription (IVT) reactions utilizing PCR amplified templates encoding the four Yamanaka reprogramming factors (Oct4, Sox2, Klf4, and c-Myc) with or without the addition of Lin28.

To decrease the immunogenicity of transfected sRNA (mediated by RIG-I, PKR, TLR7, and TLR8), the IVT reactions were carried out in the presence of unmodified and modified ribonucleotides (5-methylcytidine and pseudouridine) and an antireverse diguanosine cap analog. The IVT reaction products were then DNase treated to remove template DNA and phosphatase treated to remove 5'triphosphates from residual uncapped synthetic mRNA. The protein expression from these highly stable synthetically produced mRNAs reached its peak within 12 h of transfection and significantly decreased thereafter. Therefore, for the continuous high protein expression required for reprogramming daily mRNA transfection is required for up to 18 days.

The efficiency of synthetic mRNA reprogramming ranges from 0.6% to 4.4%. The highest efficiency of reprogramming was obtained using transfection of all five factors (Oct4, Sox2, Klf4, c-Myc, and Lin 28) in hypoxic (5% O_2) culture conditions. mRNA reprogramming efficiency was compared to traditional retroviral reprogramming and was found to be increased 0.04% versus 1.4%, respectively, and proceed with faster kinetics (ES like colonies appeared in 13 – 15 days vs. 25 – 29 days respectively).

Reprogramming of human cells with modified synthetic mRNAs is an efficient process that results in transgene free iPSCs, but it is not without its pitfalls. Specifically, mRNA mediated reprogramming is a complex process requiring multiple steps (mRNA production, puri-fication, transfection, etc.), numerous quality control measures and daily mRNA transfections to maintain high protein expression. Additionally, mRNA reprogramming may not benefit from the addition of small molecules as much as traditional viral reprogramming, as demonstrated by limited increases in efficiency when valproic acid (VPA) was included in reprogramming experiments.

3.2.2 MicroRNAs

For years it has been known that microRNAs (miRNAs), small noncoding RNAs that are thought to fine tune protein expression via binding with the 3'UTR of mRNA, play a role in maintaining pluripotency. It has also been shown that incorporation of miRNAs into traditional reprogramming strategies results in an increased reprogramming efficiency. The true impact that miRNAs can have on reprogramming was demonstrated in 2011 when Miyoshi et al. generated iPSCs from mice and human somatic cells using only miRNAs[23]. Although previous studies had successfully reprogrammed mouse and human somatic cells using miRNA, these miRNAs were expressed from integrated viral vectors. miRNA mediated reprogramming was accomplished via transfection of a cocktail of three mature miRNAs: miR-200c, miR-302s, and miR-369s. This miRNA cocktail was used to successfully reprogram mouse adipose stromal cells (mASCs), mouse embryonic fibroblasts (mEFs), human adipose stromal cells (hASCs) and human dermal fibroblasts (hDFs).

This reprogramming protocol calls for four serial transfections of the miRNA reprogramming cocktail every 48 h. After the eighth day, transfection cycle cells are transferred to ESC culture conditions and ESC like colonies begin to appear by 20 days after the first transfection. The efficiency of miRNA mediated reprogramming of mASCs is similar to the original retroviral mediated reprogramming of mEFs at 0.01% and, as one might expect, reprogramming of human cells is less efficient with a frequency of two iPSC colonies per 1×10^5 starting cells (0.002%).

Similar to other non-DNA mediated reprogramming strategies, the miRNA technique requires no screening to exclude exogenous DNA sequences being incorporated into iPSC clones. Additionally, the reprogramming efficiency is not substantially decreased as compared to other non-integrative methods. Although multiple miRNAs and multiple transfections are required to reprogram cells, Miyoshi et al. was able to efficiently induce pluripotency without the use of small molecules.

3.2.3 Recombinant Proteins

Similar to transfection of modified synthetic mRNAs, the introduction of recombinant reprogramming proteins themselves is sufficient to induce pluripotency.

In an attempt to develop transgene free iPSCs for their potential clinical use, Kim et al. and Zhou et al. both developed systems where recombinant reprogramming proteins (OSKM) tagged with a polyarginine cell penetrating peptide (CPP) could be used, with or without the addition of VPA, to reprogram mEFs and human newborn fibroblasts (hNFs) respectively. These two protocols differ in that one method uses bacterial protein expression followed by isolation, solubilization, refolding and purification steps and the other uses a total cell extract isolated from eukaryotic cells expressing the CPP tagged reprogramming factors[24-25].

In Kim et al. 's approach, mEFs were transduced with bacterially expressed recombinant proteins in the presence of VPA four times over a 9-day period. After 21 – 26 days of protein transduction, ESC like colonies were picked and further expanded.

Similar to other methods of reprogramming, this protocol is also relatively inefficient. Even in the presence of valproic acid, a HDAC inhibitor shown to increase reprogramming efficiency about 100-fold, only three iPSC clones were isolated from 5×10^4 starting cells, an efficiency of 0.006%.

Zhou et al. 's approach used retrovirally transduced HEK293 cells to express the CPP tagged reprogramming factors. Total cell extracts isolated from reprogramming factor expressing HEK293 cells were then used to reprogram hNFs in the absence of small molecules. In order to avoid cytotoxicity associated with prolonged exposure to the total cell extract, the authors used multiple cycles of 16 h protein treatment followed by 6 days of culture in ESC media.

A minimum of six cycles was required before alkaline phosphatase positive cells could be detected, a total of 42 days of protein transduction. ESC like colonies were picked 14 days after the end of the last protein transduction cycle. Using this protocol, the authors were able to isolate five iPSC clones from 5×10^5 starting cells, an efficiency of 0.001%. Whole protein extracts were used and titration of reprogramming factors was not performed, so this protocol must still be optimized. Additionally, reprogramming was accomplished in the absence of any chemical treatments so treatment with small molecules may increase reprogramming efficiency as seen with other reprogramming protocols.

Recombinant protein mediated reprogramming does not require extensive screening to rule out the introduction of exogenous genetic material resulting from integration events or inefficient transgene removal. However, both protocols require multiple protein transductions over an extended period of time to achieve reprogramming, and neither protocol is as efficient as the modified mRNA approach.

4 Adult Neural Stem Cells

Individualized therapy using adult stem cells constitutes a revolutionary vision for molecular medicine of the future. It now appears feasible to treat an individual patient's disease with native or modified stem cells collected from the same patient.

Neurodegenerative disease is a high-priority goal for stem cell therapy due to the tremendous clinical urgency to reduce the worldwide suffering associated with this class of diseases. Studies of the origin and function of neural stem cells reveals that the adult brain can generate new neurons. This finding provides the rationale for the therapeutic application of adult neural stem cells to treat neuronal damage or loss.

4.1 Neural Stem Cells: Origins and Functions

Neural stem cells comprise the subset of multipotent cells which gives rise to neurons and glial cells. Through asymmetric divisions coordinated by their microenvironment niches, these cells give rise to the most complex of all structures, the human brain (Fig. 2 – 1).

Neural stem cells are first identifiable in the developing embryo and are thought to function in differing capacities throughout the life of the individual[26]. All areas of neurogenesis and neural repair are of interest to researchers, but an area of particular medical importance is how these phenomena function in the adult. Many neurodegenerative conditions, such as Parkinson's disease, amyotrophic lateral sclerosis (ALS), stroke, or spinal cord injury, are especially devastating because we currently have no way of replacing the brain or cord cells they destroy. Recent research, though, suggests that stem cell therapy might contribute significantly to future treatment of these disorders[27].

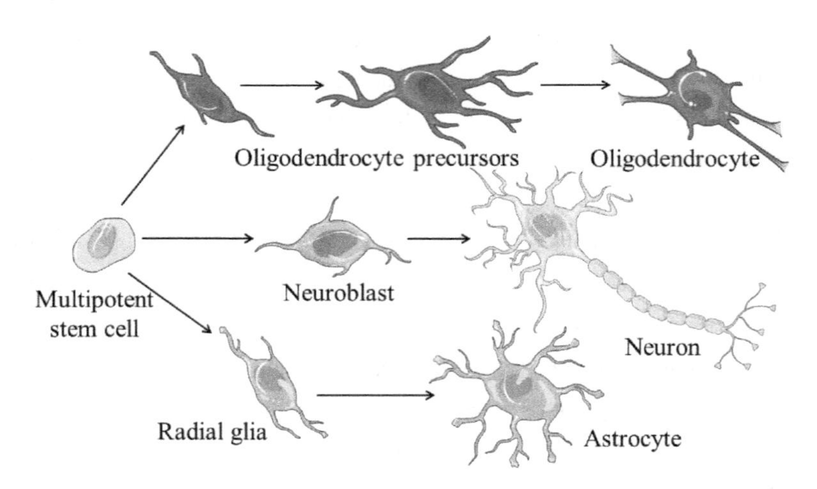

Fig. 2 – 1 Central nervous system cell lines

Illustrated by 曾宇晴.

Reference:

BAKHSHINYAN D, ADILE A A, QAZI M A, et al. Introduction to cancer stem cells: past, present, and future [J]. Methods in molecular biology (Clifton, NJ). 2018, 1692: 1 – 16.

There are several drawbacks to the use of hESCs for therapeutic purposes. Many individuals have ethical qualms about the use of cells derived from human embryos. Furthermore, implantation of ESCs has frequently led to the formation of teratomas, cancers containing cells of all three germ layers. Because of the concerns about ESCs, many investigators are evaluating adult neural stem cells. Because we know that adult brains are naturally capable of cellular regeneration, scientists pursuing cures to neurodegenerative diseases might be able to sidestep many of ethical, legal, and practical quandaries that accompany the use of embryonic or fetal stem cells.

During neurogenesis, the multipotent stem cells of the brain give rise to lineage-restricted cells which themselves are progenitors. They in turn give rise to neuronal and neuroglial cells.

4.2 Adult Brains Can Generate New Neurons

The first indication that neural stem cells might persist in adult mammals came in the 1960s with a series of studies conducted by Joseph Altman and others[28]. In the first study, Altman and Das injected ^{3}H-labeled thymidine into adult rats. The tritiated nucleoside is incorporated into the daughter strands of DNA during the S phase of cell division, and can thus be used to identify newly generated nuclei. Autoradiography revealed cell division occurring in the rat hippocampus. However, the interpretation that this suggested the development of neurons was challenged by some biologists on the basis that the labeled cells were indistinguishable from glial cells using light microscopy[29-30].

A later study by Kaplan and Hinds sought to establish with certainty the fates of cells that resulted from divisions in the post-adolescent brain. Adult rats were injected with ^{3}H-thymidine and allowed to live for 30 days before being sacrificed. Various regions of the brain were then sectioned at 1 μm thickness and mounted to slides. If a labeled cell displayed the morphological characteristics of a neuron, specifically dentritic processes, investigators further cut and mounted a thin section of this slide. The new slides were then observed using electron microscopy, and they revealed labeled newborn cells in the hippocampus, the olfactory bulb, and the neocortex. The greater resolution of the electron microscope allowed for a more accurate identification of morphological features; these cells were seen to possess uniquely neuronal characteristics, including dendrites and Gray's type 2 synapses[29].

These studies demonstrated that some mammalian brains are capable of generating new neurons even in adulthood. However, initial experiments using animals more closely related to humans, such as monkeys, yielded discouraging results. Pasko Rakic was the first to investigate primates by Altman's method of radiolabeling DNA[31]. In his study, postadolescent rhesus monkeys were injected with tritiated thymidine and sacrificed at varying intervals after treatment. Some animals were killed only after 3 days to prevent the possibility that neurons might divide numerous times and essentially wash out the ^{3}H-thymidine concentration in their DNA.

Other monkeys, though, were permitted to live an additional 6 years after injection; the stated purpose of this long interval was to give maximal time for the appearance of any slow-developing neuronal characteristics. Rakic analyzed all major structures of the brain by dark field illumination and Nomarski differential interference contrast illumination, yet he found no indication of newly formed cells with neuronal traits[31]. Additionally, while oligodendrocytes were observed in younger animals, they were largely absent from older individuals. These findings had the effect of greatly discouraging further investigation into adult neurogenesis for the next several years[30].

During this time, researchers continued to investigate a broad range of stem cells, but the idea that neurogenesis could occur in normal human adults was virtually without support until the late 1990s. Adult primate neurogenesis was first observed *in vivo* by Gould et al. in 1998[32]. Adult marmoset monkeys were administered bromodeoxyuridine (BrdU) intravenously. BrdU is a thymidine analog and so is incorporated into DNA during replication.

The animals were killed at either 2 h or 3 weeks, and their brains were sectioned and analyzed immunohistochemically. Researchers screened for both BrdU and neuron-specific enolase (NSE), a glycolytic enzyme not expressed in non-neuronal cells in primates.

Many labeled cells were observed in the dentate gyri of monkeys killed at 2 h, and more than 80% of these possessed the morphological traits of granule cell precursors. Likewise, many BrdU labeled nuclei were found in monkeys killed at 3 weeks. More than 80% of these had morphological characteristics of terminally differentiated granule neurons. This experiment presented the first direct evidence of *in vivo* adult neurogenesis in primates.

4.3 Adult Neural Stem Cells: Implications for Therapy

In the last decade, evidence has been gleaned from models that neurogenesis occurs predominantly in two regions of the adult human brain: the subgranular zone (SGZ) of the hippocampal dentate gyrus and the subventricular zone (SVZ) /olfactory bulb[33-34]. Despite having relatively few and small neurogenic sites, the human central nervous system is now believed to possess neural precursor cells throughout the brain and spinal cord. Corresponding cells in rodents and higher mammals can be translocated to neurogenic areas and incited to differentiate. Additionally, direct studies of humans, though less frequent, have provided even more certain data, eliminating the possible disconnect between model and human physiologies.

If certain areas of the human brain are neurogenic, it is at least theoretically possible for us to artificially stimulate the genesis of new neurons. If neural tissue is able to incorporate newly formed or newly introduced cells, we may be able to minimize or even eradicate the deleterious effects of maladies such as Parkinson's disease, stroke, ALS, and spinal cord injury. For example, neural cell therapy may be used in several ways to promote axonal regeneration after spinal cord injury:

(1) As a chemoattractant at the site of injury allowing trophic factors to accumulate thus enhancing regenerative capacity.

(2) As a scaffold for regeneration of axonal tissue.

(3) To replace damaged/dead cells.

Therapies relying on embryonic and fetal stem cells have been hindered by the associated ethical, political, and availability concerns. Although many of these studies were not concerned with adult stem cells, they have laid groundwork for the effective pursuit of regenerative therapy, in addition to their own direct contributions to medicine[35].

5 Experimental Method

Tissue engineering is an interdisciplinary field that applies the principles and methods of bioengineering, material science, and life sciences toward the assembly of biological substitutes that will restore, maintain, and improve tissue functions or a whole organ following damage either by disease or traumatic processes.

The general principle of tissue engineering lies on four factors: scaffold [three-dimensional (3D) tissue structures that guide the organization, growth, and differentiation of cells], ECM (that can provide the optimal conditions for cell adhesion, growth and differentiation, by controlling environmental factors), growth factors (soluble peptides capable of binding cellular receptors and producing permissive or preventive cellular response towards differentiation and/or proliferation of cells), and cells (viable cells, that are non-immunogenic, highly proliferative, easy to harvest, and pluripotent, with the ability to differentiate into a variety of cell types with specialized functions).

The main success in the field of *in vitro* tissue engineering has come from the use of primary cells. However, this strategy has limitations, because of the invasive nature of cell collection and the potential for cells to be in a diseased state. Therefore, attention has become focused upon the use of stem cells, including ESCs, bone marrow-mesenchymal stem cells (BM-MSCs), and umbilical cord-derived mesenchymal stem cells (UC-MSCs). Feeder cells provide conductive environments for growth of many stem cell lines. However, there remains the chance of contamination with xenogenic materials, and to prevent that, feeder-free systems for growing stem cells are being developed.

Also known as human pluripotent stem cells, human embryonic stem cells (hESCs) are derived from human embryos or human fetal tissues. They are self-replicating, and are also known to develop into cells and tissues of the three primary germ layers called the ectoderm, mesoderm, and endoderm (the primary layers of cells in the embryo from which all tissues and organs develop). hESCs have been used in a tissue engineering format to induce differentiation and amplification into the desired type of cell. In this section, we show how to induce hESCs to differentiate into the non-ciliated squamous epithelial cells present in the lung[36].

MSCs are multipotent stromal cells that have the ability to differentiate into a variety of cell types, including osteoblasts or bone cells, chondrocytes or cartilage cells, and adipocytes or fat cells. MSCs do not differentiate into hematopoietic cells. They can be derived from bone marrow or other non-marrow tissues, such as adipose tissue, adult muscle, corneal stroma, or dental pulp of baby. MSCs have the capability to regenerate different tissues, but not the capacity to reconstitute entire organs. The most primitive form of MSC can be isolated from the umbilical cord (UC) tissue, namely Wharton jelly, and the umbilical cord blood (UCB). Wharton jelly contains a higher concentration of MSCs than UCB, whereas the UCB is a source of hematopoietic stem cells. The UCB MSCs have more primitive properties than adult MSCs obtained later in life, which makes them a good source of MSCs for clinical application.

Dendritic cells (DCs) are antigen-presenting cells (APCs) which play a critical role in the regulation of the adaptive immune response. Also referred to as professional APCs, they have the ability to induce a primary immune response in resting, naive T lymphocytes. DCs are capable of capturing antigens, processing them, and presenting them on the cell surface along with appropriate co-stimulation molecules. DCs can be isolated from mouse spleen and other mouse tissues, as well as from human tissues[37].

Methods of isolating and culturing various types of stem cells, like hESCs, human umbilical cord derived mesenchymal stem cells (hUC-MSCs), murine bone marrow derived mesenchymal stem cells (mBM-MSCs), and murine adipose tissue derived mesenchymal stem cells (mAD-MSCs) are described below[12-13, 16, 38-39].

5.1 Materials

5.1.1 Cells

(1) Undifferentiated hESCs.

(2) MSCs isolated from mBM-MSCs, mAD-MSCs, and hUC-MSCs.

5.1.2 Culture Medium

(1) ESC conditioned medium: knockout Dulbecco's modified Eagle medium (KO-DMEM), 20% knockout serum replacement (KOSR), 1 mM sodium pyruvate, 0.1 mM β-mercaptoethanol (β-ME), 0.1 mM minimum essential medium (MEM), 1% nonessential amino acids (NEAA), 1 mM L-glutamine, and 2 ng/mL basic fibroblast growth factor (bFGF).

(2) Embryoid body (EB) medium: KO-DMEM, 20% KOSR, 20% non-heat inactivated fetal calf serum (FCS), 1% NEAA, 1 mM L-glutamine, and 0.1 mM β-ME.

(3) Small airways growth medium (SAGM): small airways basal medium, 30 μg/mL bovine pituitary extract, 5 μg/mL insulin, 0.5 μg/mL hydrocortisone, 0.5 μg/mL gentamicin sulfate-amphotericin B, 0.5 mg/mL bovine serum albumin (BSA), 10 μg/mL transferrin, 0.5 μg/mL epinephrine, and 0.5 ng/mL recombinant human epidermal growth factor (rhEGF).

(4) Bronchiolar epithelium growth medium (BEGM): bronchiolar epithelial basal medium, 30 μg/mL bovine pituitary extract, 5 μg/mL insulin, 0.5 μg/mL hydrocortisone, 0.5 μg/mL gentamicin sulfate-amphotericin B, 0.1 ng/mL retinoic acid, 10 μg/mL transferrin, 6.5 ng/mL triiodothyronine, 0.5 μg/mL epinephrine, and 0.5 ng/mL rhEGF.

(5) Growth medium: DMEM, 10% fetal bovine serum (FBS), 1% penicillin-streptomycin (Pen-Strep).

(6) Freezing medium: 90% DMEM + FBS, 10% dimethyl sulfoxide (DMSO).

(7) Adipose tissue derived MSC (AD-MSC) culture medium: DMEM, 20% FBS.

5.1.3 Buffers and Reagents

(1) Dispase solution: 1.2 U/mL dispase dissolved in Ca^{2+} and Mg^{2+}-free phosphate buffered saline (PBS), 10% ESC qualified FBS.

(2) DMSO neutralizing medium: DMEM, 10% FBS.

(3) Cell detachment solution (pH 8.0): 0.05% trypsin, 0.5 mM EDTA.

(4) Adipose tissue digestion solution: 1 × PBS, 2% BSA, 2 mg/mL collagenase A.

(5) Recombinant murine granulocyte macrophage colony stimulating factor (rm GM-CSF): 1 000 ng/mL rm GM-CSF stock.

(6) 1 × PBS (pH 7.4): 137 mM NaCl, 2.7 mM KCl, 10 mM Na_2HPO_4, 1.8 mM KH_2PO_4.

(7) Hanks' balanced salt solution (HBSS) (pH 7.0): 1.26 mM $CaCl_2$, 5.33 mM KCl, 0.44 mM KH_2PO_4, 0.50 mM $MgCl_2 \cdot 6H_2O$, 0.41 mM $MgSO_4 \cdot 7H_2O$, 138 mM NaCl, 4 mM $NaHCO_3$, 0.30 mM Na_2HPO_4, 5.60 mM glucose, 0.03 mM phenol red.

5.1.4 Instruments

(1) Humidified CO_2 incubator, at 37 ℃, 5% CO_2.

(2) Biosafety cabinet.

(3) Centrifuge.

(4) Centrifuge tubes.

(5) Tissue culture plates (10 mm, 60 mm, 90 mm).

(6) 40 μm nylon mesh filter.

5.2 Methods

5.2.1 Culture and Differentiation of hESCs

(1) Expansion of hESCs:

A. Plate primary mEF feeder cells, prepared from timed pregnant CF-1 female mice (day 13.5 of gestation), in conditioned medium, in 6-well 10 mm tissue culture plates [see Note (1)].

B. Incubate the cells in a humidified 5% CO_2 incubator at 37 ℃.

C. After growth, γ-irradiate the mEF cells with 3 000 rads for 5 min, to stop differentiation of mEF cells (see Note 2).

D. Plate the hESCs on the γ-irradiated mEF feeder cells in ESC conditioned medium, and incubate in a humidified 5% CO_2 incubator at 37 ℃.

(2) Embryoid Body Formation:

A. Treat colonies with well-defined boundaries and minimum differentiation, with dispase solution at 37 ℃, till the ESC colonies nearly detach from the plates.

B. Wash the colonies off the plates, then wash twice with ESC conditioned medium without bFGF.

C. Resuspend the cells in EB medium, and transfer to ultra-low attachment tissue culture plates.

D. Grow for 4 days at 37 ℃.

(3) Generation of non-ciliated pulmonary epithelial cells:

A. Transfer EBs to tissue culture plates after dispase digestion.

B. Culture EBs for 12 days in SAGM or BEGM (Refresh medium every other day).

C. From the day 12 culture in SAGM or BEGM, flow sort the alveolar epithelium cells on the basis on surface expression of SP-C and AQP-5.

D. Grow the flow sorted SP-C^+ and $AQP5^+$ cells in SAGM or BEGM for 4 days.

5.2.2 Isolation and Culture of Human Umbilical Cord MSCs

(1) Collect human UC from hospitals (see Note 3), and store in DMSO at 80 ℃, in 50 mL centrifuge tubes.

(2) On the day of the experiment, thaw the tube to 37 ℃ quickly, till only a few ice crystals remain.

(3) Wash the cord in DMSO-neutralizing medium and transfer to petri dish containing 7 mL of the growth medium.

(4) Squeeze out the inner contents of the cord into the medium using forceps and scalpel.

(5) Chop the remaining cord into small sections using surgical blades and place them in 20 mL growth medium in a 90 mm tissue culture plate [see Note (1)].

(6) Add the inner contents of the cord into the plate containing the sections and incubate at 37 ℃, with 5% CO_2 for 5 days. Keep undisturbed for 3 days, then check under microscope for adherent cells. Continue incubation till 5 days.

(7) Once cells become confluent, they can be passaged, after detachment with trypsin-EDTA, or frozen and stored in cryovials, containing freezing medium.

5.2.3 Isolation of Murine Bone Marrow Derived MSCs

(1) Collect femur and tibia from mice in growth medium, and remove the flesh and muscles [see Notes (4) and (5)].

(2) Under a biosafety cabinet, cut the ends of the bones and flush the contents of the bone with growth medium, into sterile centrifuge tubes.

(3) Plate 5×10^4 cells in 60 mm tissue culture plates [see Note (1)], in growth medium, for 3 h at 37 ℃, with 5% CO_2.

(4) Remove non-adherent cells carefully after 3 h and add fresh medium.

(5) When the cells become confluent, treat with 0.5 mL of 0.25% trypsin-EDTA for 2 min at room temperature (25 ℃), to get a purified population of BM-MSCs.

5.2.4 Isolation of Murine Adipose Tissue Derived MSCs

(1) Isolate inguinal adipose tissue from 12 to 14-week-old BALB/c mice [see Notes (4) and (5)].

(2) Digest in adipose tissue digestion solution for 15 – 20 min.

(3) Filter through 40 μm nylon filter mesh, centrifuge at 500 × g for 5 min, and resuspend pellet in AD-MSC growth medium.

(4) Plate 5×10^4 cells in 60 mm tissue culture plates [see Note (1)] and incubate at 37 ℃ with 5% CO_2.

(5) Change medium every 2 days, and passage after cells reach 80% – 90% confluence.

5.2.5 Notes

(1) Treat all tissue culture plates with 0.1% gelatin for better adherence of cells. Treat plates with gelatin overnight at room temperature. Discard the gelatin, air-dry the plates in a laminar airflow hood for 45 min, and then expose them to UV for 30 min.

(2) To stop differentiation of mEF feeder cells, mitomycin C (at a final concentration of 10 μg/mL) for 2 – 3 h at 37 ℃ can be used instead of γ-irradiation.

(3) Collection of human UC should be done with the consent of the patients.

(4) Animal sacrifices should be done ethically.

(5) All animal handling should be done with proper protective gear, and under the laminar airflow hood.

Supplement

List of Abbreviations	
AAV	adeno-associated virus
ASCs	adipose-derived stem cells
ECM	extracellular matrix
ESCs	embryonic stem cells
FACS	fluorescent activated cell sorter
FBS	fetal bovine serum
hESCs	human embryonic stem cells
HSCs	hematopoietic stem cells
iPSCs	induced pluripotent stem cells
LVV	lentiviral vector
mEF	mouse embryonic fibroblast
MSCs	mesenchymal stem cells
shRNA	short hairpin RNA
siRNA	small interfering RNA
ZFN	zincfinger nuclease

Key Words List	
丙戊酸	valproic
成骨的	osteogenic
成软骨的	chondrogenic
成脂的	adipogenic
抽脂	lipoaspirate
单层的	unilamellar
多谱系	multilineage
多顺反子	polycistronic
寡聚的	oligomeric
畸胎瘤	teratoma
假尿苷	pseudouridine
聚乙烯亚胺	polyethylenimine
链脲佐菌素	streptozotocin

(To be continued)

Key Words List	
链脲佐菌素	streptozotocin
内体	endosomal
千碱基	kilobases
软骨形成	chondrogenesis
细胞计数	cytometry
锌指	zincfinger
新皮层	neocortex
血糖	glycemia

References

[1] EIGES R. Genetic manipulation of human embryonic stem cells [J]. Methods in molecular biology, 2016, 1307: 149 – 172.

[2] EIGES R, SCHULDINER M, DRUKKER M, et al. Establishment of human embryonic stem cell-transfected clones carrying a marker for undifferentiated cells [J]. Current biology, 2001, 11 (7): 514 – 518.

[3] HUBER I, ITZHAKI I, CASPI O, et al. Identification and selection of cardiomyocytes during human embryonic stem cell differentiation [J]. FASEB journal, 2007, 21 (10): 2551 – 2563.

[4] COSTA M, DOTTORI M, SOURRIS K, et al. A method for genetic modification of human embryonic stem cells using electroporation [J]. Nature protocols, 2007, 2 (4): 792 – 796.

[5] GUO G, HUANG Y, HUMPHREYS P, et al. A PiggyBac-based recessive screening method to identify pluripotency regulators [J]. PLOS one, 2011, 6 (4): e18189.

[6] URBACH A, SCHULDINER M, BENVENISTY N. Modeling for Lesch-Nyhan disease by gene targeting in human embryonic stem cells [J]. Stem cells, 2004, 22 (4): 635 – 641.

[7] ZWAKA T P, THOMSON J A. Homologous recombination in human embryonic stem cells [J]. Nature biotechnology, 2003, 21 (3): 319 – 321.

[8] LI M, SUZUKI K, KIM N Y, et al. A cut above the rest: targeted genome editing technologies in human pluripotent stem cells [J]. The journal of biological chemistry, 2014, 289 (8): 4594 – 4599.

[9] SMITH-ARICA J R, THOMSON A J, ANSELL R, et al. Infection efficiency of human and mouse embryonic stem cells using adenoviral and adeno- associated viral vectors

[J]. Cloning stem cells, 2003, 5 (1): 51-62.

[10] PARK J H, KIM S J, OH E J, et al. Establishment and maintenance of human embryonic stem cells on STO, a permanently growing cell line [J]. Biology of reproduction, 2003, 69 (6): 2007-2014.

[11] SHERMAN L S, CONDÉ-GREEN A, KOTAMARTI V S, et al. Enzyme-free isolation of adipose-derived mesenchymal stem cells [J]. Methods in molecular biology, 2018, 1842: 203-206.

[12] FERRIN I, BELOQUI I, ZABALETA L, et al. Isolation, culture, and expansion of mesenchymal stem cells [J]. Methods in molecular biology, 2017, 1590: 177-190.

[13] PRAT M, OLTOLINA F, ANTONINI S, et al. Isolation of stromal stem cells from adipose tissue [J]. Methods in molecular biology, 2017, 1553: 169-182.

[14] WILSON A, CHEE M, BUTLER P, et al. Isolation and characterisation of human adipose-derived stem cells [J]. Methods in molecular biology, 2019, 1899: 3-13.

[15] RIGOTTI G, MARCHI A, GALIÈ M, et al. Clinical treatment of radiotherapy tissue damage by lipoaspirate transplant: a healing process mediated by adipose-derived adult stem cells [J]. Plastic and reconstructive surgery, 2007, 119 (5): 1409-1422.

[16] MAHMOUDIFAR N, DORAN P M. Mesenchymal stem cells derived from human adipose tissue [J]. Methods in molecular biology, 2015, 1340: 53-64.

[17] GARDNER O F, ALINI M, STODDART M J. Mesenchymal stem cells derived from human bone marrow [J]. Methods in molecular biology, 2015, 1340: 41-52.

[18] THOMSON J A, ITSKOVITZ-ELDOR J, SHAPIRO S S, et al. Embryonic stem cell lines derived from human blastocysts [J]. Science, 1998, 282 (5391): 1145-1147.

[19] TAKAHASHI K, YAMANAKA S. Induction of pluripotent stem cells from mouse embryonic and adult fibroblast cultures by defined factors [J]. Cell, 2006, 126 (4): 663-676.

[20] YU J, VODYANIK M A, SMUGA-OTTO K, et al. Induced pluripotent stem cell lines derived from human somatic cells [J]. Science, 2007, 318 (5858): 1917-1920.

[21] OKITA K, ICHISAKA T, YAMANAKA S. Generation of germline-competent induced pluripotent stem cells [J]. Nature, 2007, 448 (7151): 313-317.

[22] HAYES M, ZAVAZAVA N. Strategies to generate induced pluripotent stem cells [J]. Methods in molecular biology, 2013, 1029: 77-92.

[23] MIYOSHI N, ISHII H, NAGANO H, et al. Reprogramming of mouse and human cells to pluripotency using mature microRNAs [J]. Cell stem cell, 2011, 8 (6): 633-638.

[24] KIM D, KIM C H, MOON J I, et al. Generation of human induced pluripotent stem cells by direct delivery of reprogramming proteins [J]. Cell stem cell, 2009, 4 (6):

472 -476.

[25] ZHOU H, WU S, JOO J Y, et al. Generation of induced pluripotent stem cells using recombinant proteins [J]. Cell stem cell, 2009, 4 (5): 381 -384.

[26] GAGE F H, RAY J, FISHER L J. Isolation, characterization, and use of stem cells from the CNS [J]. Annual review of neuroscience, 1995, 18: 159 -192.

[27] ZENG X, RAO M S. Human embryonic stem cells: long term stability, absence of senescence and a potential cell source for neural replacement [J]. Neuroscience, 2007, 145 (4): 1348 -1358.

[28] ALTMAN J, DAS G D. Autoradiographic and histological studies of postnatal neurogenesis. I. A longitudinal investigation of the kinetics, migration and transformation of cells incorporating tritiated thymidine in neonate rats, with special reference to postnatal neurogenesis in some brain regions [J]. The journal of comparative neurology, 1966, 126 (3): 337 -389.

[29] KAPLAN M S, HINDS J W. Neurogenesis in the adult rat: electron microscopic analysis of light radioautographs [J]. Science, 1977, 197 (4308): 1092 -1094.

[30] GOULD E. How widespread is adult neurogenesis in mammals? [J]. Nature reviews neuroscience, 2007, 8 (6): 481 -488.

[31] RAKIC P. Limits of neurogenesis in primates [J]. Science, 1985, 227 (4690): 1054 -1056.

[32] GOULD E, TANAPAT P, MCEWEN BS, et al. Proliferation of granule cell precursors in the dentate gyrus of adult monkeys is diminished by stress [J]. Proceedings of the National Academy of Sciences of the United States of America, 1998, 95 (6): 3168 -3171.

[33] LOIS C, ALVAREZ-BUYLLA A. Proliferating subventricular zone cells in the adult mammalian forebrain can differentiate into neurons and glia [J]. Proceedings of the National Academy of Sciences of the United States of America, 1993, 90 (5): 2074 -2077.

[34] GRITTI A, BONFANTI L, DOETSCH F, et al. Multipotent neural stem cells reside into the rostral extension and olfactory bulb of adult rodents [J]. The journal of neuroscience, 2002, 22 (2): 437 -445.

[35] LETCHER J M, COX D N. Adult neural stem cells: isolation and propagation [J]. Methods in molecular biology, 2012, 823: 279 -293.

[36] BANERJEE E R. Looking for the elusive lung stem cell niche [J]. Translational respiratory medicine, 2014, 2: 7.

[37] KAR S, MITRA S, BANERJEE E R. Isolation and culture of embryonic stem cells, mesenchymal stem cells, and dendritic cells from humans and mice [J]. Methods in

molecular biology, 2016, 1516: 145 - 152.

[38] ODABAS S, ELÇIN A E, ELÇIN Y M. Isolation and characterization of mesenchymal stem cells [J]. Methods in molecular biology, 2014, 1109: 47 - 63.

[39] LIU G, CHEN X. Isolating and characterizing adipose-derived stem cells [J]. Methods in molecular biology, 2018, 1842: 193 - 201.

Chapter 3 Stem Cells and Human Diseases
第三章 干细胞与人类疾病

［中文导读］

近几十年来，医疗技术快速发展，对人类健康做出了巨大贡献，但人类至今仍缺乏应对癌症、艾滋病和糖尿病等重大疾病的有效手段。20 世纪末以来，生物医药的研发出现了两个新的特征：一方面，新药开发日趋困难。美国食品药品管理局批准的新药数量由 20 世纪 90 年代中期的每年 50 余个降至近年来的每年约 20 个，新化学结构的药物研究难度越来越大，而上市的生物药物数量也未见明显的增长。另一方面，分离和培养第一个人多能干细胞系的成功，予以医学界极大的惊喜，其在生物医学的研究上，开辟了治疗疾病新的疆界。许多疾病及功能失调往往是由于细胞功能障碍或组织被破坏。如今，一些捐赠的器官和组织常常用以取代生病的或遭破坏的器官或组织，而受这些疾病折磨的患者数量远远超过了可供移植的器官数量。在干细胞疗法中，一个重要的步骤是选择合适的干细胞来源。干细胞经适当刺激后可发展为特化的细胞，使替代细胞和组织来源的更新成为可能，从而可用于治疗无数的疾病，包括帕金森病、阿尔茨海默病、脊髓损伤、心脏病、糖尿病、骨关节炎等。

1999 年，有关干细胞的研究被美国 *Science* 杂志评为“年度十大科学进展”之首；2000 年，干细胞研究再次被评为“年度十大科学进展”之一。此后，干细胞研究快速发展。近几年来，诱导多能干细胞（induced pluripotent stem cells）的重大进展进一步将干细胞研究推向了生物医药领域的前沿。干细胞治疗研究被认为具有重大的科学意义、社会意义和广阔的产业前景。

干细胞疗法为许多患者带来了希望，科学界和医学界仍在努力进一步阐明干细胞生物学的复杂性，并提供证据以支持这些药物用于治疗疾病的合理性。人类干细胞系的发展值得人们开展严密的科学考察，对新的疗法进行评估，对预防策略和伦理问题进行公开讨论。

1 Stem Cell Therapy for Neurological Disorders

Neurological disease encompasses a diverse group of disorders of the central and peripheral nervous systems. The scope of treatment options for neurological disease is limited, and drug approval rates for improved treatments remain poor when compared with other therapeutic areas. Many forms of stem cell therapy exist, including the use of neural, hematopoietic and

MSCs. Cell therapies derived from differentiated ESCs and iPSCs are also starting to feature prominently. Over 200 clinical studies applying various stem cell approaches to treat neurological disease have been registered to date, the majority of which are for multiple sclerosis, stroke, and spinal cord injuries.

Stem cell therapy for neurological disorders in principle, for a stem cell therapy to be successful in treating diseases of the brain where neurons are damaged or defective, the treatment should aim to repair, replace or at least prevent future deterioration. The goal of stem cell therapy is thus to enable localisation of therapeutic cells to impaired/injured regions of the brain, to stimulate tissue repair and maintenance via a paracrine effect, and potentially even to generate new neurons. Each of the described stem cell types is being applied in the treatment of neurological disease, for which a few advantages and disadvantages are listed in Tab. 3 – 1.

Tab. 3 – 1 Advantages and disadvantages of the different stem cell types

Stem cells	Advantages	Disadvantages
NSCs	Prototype stem cells for treating neurological disease	Limited resource with ethical implications around procurement
	NSC-like cells can be derived from other stem cell types (pluripotent and multipotent)	Poorly understood stem cell biology and least explored in clinical studies
	—	Tumorigenic risks if derived from pluripotent stem cells
HSCs	Globally accepted form of treatment for hematological conditions	Limited experience for use in neurological disease
	Well-established industry for harvesting and preparation of clinical grade treatments	Generally limited for use as an autologous therapy (requires genetic matching of the donor and recipient if used as an allogeneic treatment); Poorly understood mechanism of action for treating certain neurological conditions
MSCs	Readily accessible resource and easily procured	Exploited by unregulated clinics globally
	No need for genetic matching	Poorly understood mechanism of action for treating certain neurological conditions
	Most likely stem cell therapy to evolve into an off-the-shelf allogenic product	—

With recent advances of stem cell biology, it is possible to differentiate pluripotent stem cells (both ESCs and iPSCs) into neural progenitor or NSC-like cells for therapeutic purposes. Protocols for differentiating bone marrow-derived MSCs into NSC-like cells have also been developed and applied for clinical application in patients with multiple sclerosis. Given their high proliferative capacity and risk for tumour formation, pluripotent stem cells are not used directly for treatment purposes, but rather are differentiated into the desired fully differentiated cell type. HSCs and MSCs are also being explored in clinical studies to investigate their potential for treating various neurological disorders[1].

1.1 Neural Stem Cells

NSCs are multipotent progenitors which have the ability to produce neurotrophic factors and/or differentiate into committed cell lineages of the central and peripheral nervous systems, including neurons and supporting glial cells (such as astrocytes and oligodendrocytes).

In the adult brain, NSCs are limited to the hippocampus and play a role in supporting plasticity. Those found in the subventricular zone (SVZ) of the hippocampus are in contact with the cerebrospinal fluid, moving radially and differentiating to young neuroblasts. NSCs in the subgranular zone (SGZ) also move radially and differentiate to neuroblasts, but are referred to as radial glia-like NSCs. Neuroblasts go on to mature and differentiate into neurons. NSCs are abundant in fetal brains, most notably at very early stages of development. Once harvested from the fetal brain, they can be dissociated and grown in the laboratory as a monolayer of cells or as floating "spheroids" that produce neurotrophic factors that can be differentiated into more defined neuronal lineages. Pluripotent stem cells are increasingly becoming an important source for the generation of neural precursor cells. Using a variety of *in vitro* conditions, cells resembling NSCs (NSC-like) can be generated from hESCs. These conditions include co-culturing with animal stromal cells (unknown stimuli), exposure to retinoic acid, and inhibition of bone morphogenetic protein (BMP) signaling. The recent development of iPSCs, which have also been shown to generate NSC-resembling cells, mitigates ethical concerns linked to the use of embryos and also introduces the possibility of autologous transplantation using patient-derived iPSCs. The rationale for therapeutic application of NSCs and NSC-like cells is based on their ability to release neurotrophic factors and to differentiate into neural and glial cells, thereby promoting neurogenesis and replacing diseased or injured areas of the brain[2].

1.1.1 Stroke

Stroke occurs either as a result of impaired blood flow (ischaemic) or rupturing of blood vessels (haemorrhagic) in the brain, resulting in diminished oxygen and nutrient supply to neurons. Preclinical studies using NSCs to treat stroke have demonstrated that, in addition to stimulating neurogenesis, NSCs are capable of releasing angiogenic factors to promote local tissue regeneration.

1.1.2 Multiple Sclerosis

Multiple sclerosis is a neurodegenerative disorder characterized by inflammatory demyelination of neurons. Recently reported findings from a Phase I clinical study (NCT01933802) showed that intrathecal administration of MSC-derived neural progenitor cells for patients with multiple sclerosis was safe and well tolerated. Although only modest improvements were observed, the initial findings warrant further clinical development and initiation of a Phase II study.

1.1.3 Parkinson's Disease

In Parkinson's disease, a neurodegenerative disorder resulting from progressive loss of dopaminergic neurons, application of NSCs with an acquired specification for a dopaminergic neuronal identity has demonstrated promising results. Dopaminergic neural progenitor cells can either be isolated from the fetal ventral midbrain or generated *in vitro* from human ESCs or iPSCs. In the single study using oocytes as a source, NSC-like cells were differentiated from human parthenogenetic stem cells obtained from chemically activated unfertilized oocytes. These cells are being used to treat patients with Parkinson's disease, with preliminary data showing safety with indications of efficacy over an ongoing 5-year period of follow-up.

1.1.4 Brain Tumors

For treatment of brain tumors, 3 registered clinical studies use fetal-derived NSCs which have been gene modified to express cytosine deaminase, an enzyme involved in the activation of chemotherapeutic agents such as 5-fluorouracil and irinotecan, which serves to localize drug activity to the site of transplantation. Data from one of these studies have been reported, wherein 15 patients were recruited and proof-of-principle for this approach was demonstrated.

1.2 Hematopoietic Stem Cells

HSCs are multipotent stem cells traditionally used for the treatment of malignant and non-malignant diseases of the blood and immune systems. They can be harvested directly from bone marrow, peripheral blood (following pre-treatment to mobilize HSCs from the bone marrow) or umbilical cord blood (UCB). Depending on the condition being treated, HSCs are either collected from the patient or from a healthy donor to be used for autologous or allogeneic transplantation, respectively. Allogeneic transplantation, however, comes with the constraint of having to genetically match the donor and recipient in order to avoid graft rejection or, more concerningly, the risk of graft versus host disease and its associated mortality. Patients are also subjected to lifelong immunosuppressive therapy.

HSCs transplantation is performed in over 80 countries and, of the more than 60 000 HSCs transplants that are done globally per annum, the vast majority (about 90%) are for treating hematological malignancies, including leukemia, lymphoma and myeloma. Secondary to these indications is the treatment of solid tumors, while non-malignant conditions include bone marrow failure, hemoglobinopathies and primary immune disorders. In all of these cases, the underlying principle of HSCs transplantation is to replenish the bone marrow with stem cells which engraft and reconstitute the immune system with a functional lineage of hematopoietic cells.

In addition to the above traditional indications, there has over the last decade been a distinct increase in the use of HSCs for regenerative purposes, particularly UCB as a source of HSCs for treating neurological conditions. There are 54 clinical studies registered to treat neurological disease with HSCs, 40 of which UCB as a source. Once harvested, an UCB unit is processed to separate mononuclear cells (via removal of the red blood cells and plasma), which ultimately comprises a high proportion of HSCs, and to a lesser extent MSCs, endothelial progenitor cells and immunosuppressive cells such as regulatory T cells and monocyte-derived suppressor cells. Altogether and once transplanted, these cells have been shown to provide a paracrine effect that ① promotes cell survival, ② stimulates proliferation and migration of NSCs, ③ induces regeneration of damaged brain cells, ④ reduces inflammation, ⑤ promotes angiogenesis. Cerebral palsy and hypoxic ischaemic encephalopathy (HIE) are the neurological indications most treated with UCB on an experimental basis. HIE is a significant risk factor for developing cerebral palsy and occurs during birth as a result of reduced blood flow to the brain. Up to 20% of these cases result in death of the newborn, and nearly 30% develop permanent neurological abnormalities. Only 1 of the 10 registered experimental clinical studies using UCB for HIE has reported data. In this pilot study with 23 participants, researchers demonstrate that the collection, preparation and infusion of fresh (non-cryopreserved) UCB cells into newborns with HIE is a feasible and safe approach.

1.3 Mesenchymal Stem Cells

MSCs, also referred to as mesenchymal stromal cells or medicinal signaling cells, have the ability to self-renew and differentiate into cells of the mesoderm, including bone, adipose and cartilage. MSCs are believed to have high treatment potential based on several unique characteristics, including ① their ability to home to a site of injury, ② their immunomodulatory and paracrine effects, ③ the fact that they are immune-privileged (i. e. do not required genetic matching), ④ that they can be procured from many sources, such as bone marrow, adipose tissue and UC Wharton's jelly. MSCs from each of these sources are, however, more prone to differentiate into cells of their origin, and are hence more committed to these lineages[3].

1.3.1 MSC Products

To date, 3 MSC products have obtained regulatory approval for patient treatment. These include Remestemcel-L (allogeneic bone marrow-derived MSCs) in Canada, New Zealand and Japan for acute graft-versus-host disease; Darvadstrocel (allogeneic adipose derived

MSCs) in Europe for fistulae in Crohn's disease; and Stemirac (autologous bone marrow-derived MSCs) in Japan for treating spinal cord injury. None of these therapies has been approved for use in the USA, nor are they being used routinely in Europe as yet. With respect to the clinical trial landscape, MSCs are being applied to treat an extraordinary number of diseases. Based on our investigations, up to 150 different indications can be identified from registered clinical trials. Amongst these are a range of neurological conditions, diabetes, stroke, osteoarthritis, emphysema, bone fractures, wounds, macular degeneration and incontinence.

1.3.2 Tissue Source

In terms of tissue source, more than half of the registered trials use MSCs derived from bone marrow, followed by adipose tissue, umbilical cord and umbilical cord blood. In each of these cases, the MSCs are expanded in cell culture to a therapeutic dose. With respect to the use of adipose tissue as a source, the treatment can either be in the form of stromal vascular fraction (SVF) or with culture-expanded MSCs, also referred to as adipose-derived stromal cells (ASCs). SVF is a concentrated extract of enzyme-digested adipose tissue which contains a mixture of ASCs, endothelial and endothelial precursor cells, lymphocytes, macrophages, smooth-muscle cells, pericytes and preadipocytes.

1.3.3 Multiple Sclerosis

For multiple sclerosis, the rationale for using MSCs is based on their immunomodulatory and neuroprotective properties. In a study reported in 2010, no adverse events and signs of clinical but not radiological efficacy were reported in 10 study participants following intravenous infusion, where it was shown that MSCs provided suggestive benefits of neuroprotection based on their immunomodulatory and anti-inflammatory properties. The most recent report using bone marrow-derived MSCs for multiple sclerosis also included follow-up administrations of MSC-conditioned medium. MSC-conditioned medium, i. e. the same cell culture medium used to expand MSCs, contains a range of cytokines, chemokines and growth factors which are postulated to further promote neuronal regeneration. In this clinical study with 10 participants, the procedure was shown to be well tolerated with relative efficacy in establishing the disease and reversing symptoms.

The future of stem cell therapy for neurological disease is promising. We expect that more clinical studies using NSC-like treatments will be registered in the future, particularly those using iPSCs as a source of differentiated cells. MSC-based treatments will remain attractive, given that they are a readily accessible resource and can be prepared with relative ease. However, therapeutic application of these cells should be based on rational premises. Early-stage clinical studies with promising indications of efficacy will progress into later-stage studies, and only those that show unequivocal efficacy in well-designed, randomized clinical trials will finally reach the market. Complex disease requires complex therapies and, although public perception is that stem cell therapy could be a magic bullet for the cure of numerous ailments, it is unlikely to be administered as a monotherapy.

1.3.4 Alzheimer's Disease

Dementia is a fatal clinical disorder characterized by amnesia, progressive cognitive impairment, disorientation, behavioral disturbance, and loss of daily function. Alzheimer's disease (AD) is the most common associated pathology. It can be argued that dementia is one of the most significant social, economic, and medical challenges of our time. Less than 5% of AD cases are familial, caused by highly penetrant autosomal mutations of the PSEN1, PSEN2, and, less frequently, the APP genes. The majority of AD cases are late onset and sporadic, with established risk factors beyond age including cardiovascular disease, low education, depression, and the apolipoprotein-E4 (ApoE4) gene. Sporadic AD is accordingly of multifactorial origins, driven in part by a complex genetic profile and in part by interacting and intersecting environmental exposures. Four core features can be discerned.

(1) Tau, an intracellular microtubule-associated protein within neurons important for structural support and axonal transport, becomes hyperphosphorylated, leading to microtubule collapse and aggregation into neurofibrillary tangles.

(2) Sequential cleavage of the APP protein by β-and γ-secretase enzymes leads to extracellular accumulation and aggregation of beta amyloid (Aβ) protein fragments, visible as amyloid plaques in the AD brain. Many pharmacological approaches have attempted to promote amyloid clearance by vaccination and decrease production via secretase inhibition. However, results from human clinical trials indicate that amyloid pathology does not correlate with clinical symptoms and therefore may not be a therapeutically relevant target.

(3) The presence of activated microglia, the resident macrophages of the central nervous system (CNS), and found in close association with amyloid plaques. Present from the early stages of the disease, their numbers then decline in the advanced AD brain. Activated microglia produce cytokines, such as tumor necrosis factor (TNF)-α, interleukin (IL)-1β, and nitric oxide (NO), which may exacerbate or attenuate neuroinflammation.

(4) Mass neuronal and synaptic loss represents the fourth core feature of AD and is the closest correlate of cognitive decline in early AD. AD-related neurodegeneration in the temporal lobe follows a distinct pattern. The entorhinal cortex is first affected, then progressing to the subiculum and CA1 hippocampal subregion and basal forebrain networks. Atrophy of these brain regions and the hippocampus overall co-vary with verbal episodic memory deficits in AD patients. In later stages of the disease neurodegeneration spreads throughout the temporal lobes, eventually affecting most cortical layers. The precise temporal sequencing of this complex admixture of pathologies in human sporadic AD is the subject of intense debate.

Due to the progressive nature of AD, if a stem cell therapy is to be successful it must target a well-defined clinical subset of patients. There is now an enormous global demand for new effective therapies that not only halt progression but also reverse symptoms. The most commonly utilized cells in recent AD studies are ESCs, MSCs, brain-derived NSCs, and iPSCs.

(1) ESCs are derived from the inner cell mass of the developing blastocyst (at embryonic day 5 to 6) and are classified as pluripotent because they possess the ability to generate cell types from the ectodermal, mesodermal, and endodermal germ layers.

(2) MSCs are involved in the development of mesenchymal tissue types and can be harvested from UCB-MSCs or Wharton's jelly, and also remain present in several adult stem cell niches including bone marrow and adipose tissue. Classified as multipotent, MSCs are able to generate multiple cell types that share a common embryonic origin, namely the mesodermal germ layer. Despite this, phenotypic expression and the differentiation potential of MSCs can vary according to the tissue of origin.

(3) NSCs are responsible for the generation of all neural cell types during development. While also present in the adult brain, they are restricted to the discrete neurogenic niches of the subventricular zone and the granular layer of the dentate gyrus in the hippocampus.

(4) iPSCs are derived from mature somatic cells *in vitro*, commonly adult dermal fibroblasts, and are genetically modified by small molecule treatment or viral vector-delivered transcription factor upregulation to become pluripotent and ESC-like in phenotype and differentiation capacity.

Endogenous repair has several theoretical approaches to the design of a stem cell therapeutic strategy for early AD. One is to target upregulation of resident NSC niches within the adult brain, in effect stimulating adult hippocampal neurogenesis to compensate for neurodegeneration. Adult hippocampal neurogenesis may have a key role in learning and memory, and so promoting this process may help counter the amnestic symptoms of early AD.

Another option has been to upregulate (pharmacologically or with gene therapy) those growth factors known to positively regulate neurogenesis, including brain-derived neurotrophic factor (BDNF), insulin growth factor-1 (IGF-1), nerve growth factor (NGF), and vascular endothelial growth factor (VEGF).

This approach is, however, complicated by several quantitative challenges. Firstly, the rate of hippocampal neurogenesis decreases with age in humans, with an estimated 800 new neurons produced daily in adulthood declining to about 100 in later life under disease-free conditions. Since the best estimates suggest neuronal number is stable in normal aging, this is therefore the minimum required to achieve neuronal equilibrium because of rapid neuronal turnover. Secondly, there is mass loss of hippocampal neurons in AD. In the dentate gyrus the loss is estimated at around 1 M, and in CA1 the loss is estimated at 5 million. Hence, to compensate for AD there would need to be an order-fold increase in hippocampal neurogenesis to normalize dentate gyrus numbers. Furthermore, adult hippocampal neurogenesis has no effect whatsoever on CA1 neurons and so the main neuronal deficit in early AD is unaddressed. Thirdly, this approach must account for the effect of AD pathology on neurogenesis, for which there is conflicting evidence from animal studies. Overall, endogenous strategies for neuronal repair in early AD lack potency and miss one of the main neuronal targets.

1.3.5 Exogenous Cell Therapy

Exogenous cell therapies aim to restore degenerate neuronal networks, and consequently cognitive function, through the introduction of stem cells. These stem cells may be used as a cellular delivery system, utilizing a paracrine "bystander" mechanism through either native or induced production of neuroprotective growth factors. Alternatively, therapeutic restoration may occur through differentiation and participation of the stem cells in repopulating degenerate neuronal circuits. This is a finely balanced, complex, and multistep process. Each class of stem cells has different propensities to achieve these approaches.

(1) ESCs. While some ESCs transplantation studies have shown a capacity to restore cognitive function in rodent models of brain injury, their clinical translation has been limited. This is in part due to their pluripotent nature, as transplantation of undifferentiated ESCs presents an inherent risk of uncontrolled cell growth and tumor formation. In vitro pre-differentiation of ESCs into NSCs circumvents some of this risk, generating predominantly cholinergic neurons and inducing improvements in spatial memory performance after transplanting into an AD rodent model.

(2) NSCs. The paracrine effect of NSCs has been shown to have significant therapeutic potential. Transplanting growth factor-secreting NSCs increased neurogenesis and cognitive function in a rodent AD model and aged primate brain, while transplantation of choline acetyltransferase overexpressing human NSCs into a cholinergic neurotoxic rodent model resulted in a reversal of spatial memory and learning deficits. While the therapeutic mechanisms behind these changes are not yet fully understood, they are likely mediated by both the paracrine release of neuroprotective or immune modulatory factors and by direct neuronal differentiation, although the widespread generation of non-neuronal glial cell types from transplanted NSCs remains a major limiting factor for neuroreplacement strategies.

(3) MSCs. Due to their accessibility, relative ease of handling, and the broad range of cell types that they are able to generate, MSCs are now among the most frequently studied stem cell type. In aged rodent models, transplanted MSCs were shown to undergo differentiation into neural cell types, increasing local concentrations of acetylcholine neurotransmitter, BDNF, and NGF, and improving locomotor and cognitive function. However, to date there has been little evidence for the functional or synaptic maturation of MSC-derived neurons *in vivo*. Moreover, genuine neuroreplacement by MSCs remains limited by low rates of neuronal differentiation and a propensity for glial cell formation *in vivo*. Potentially of greater therapeutic significance are the reported neuroprotective paracrine effects of MSCs, with the introduction of MSC-secreted factors able to stimulate proliferation, neuronal differentiation, and sur-

vival in endogenous neurogenic niches and in cellular models of AD. Similarly, in rodent AD models, MSCs transplantation has been reported to inhibit Aβ and tau related cell death, reduce Aβ deposits and plaque formation, stimulate neurogenesis, synaptogenesis, and neuronal differentiation, and rescue spatial learning and memory deficits. Some studies suggest a further anti-inflammatory and immune modulatory paracrine effect for transplanted MSCs, including upregulated neuroprotective cytokines such as IL-10, and reduced levels of pro-inflammatory cytokines TNF-α and IL-1β. Intravenously administered MSCs are also capable of crossing the blood-brain barrier and effectively migrating to regions of neural injury, without inducing a tumorigenic or immune response.

(4) iPSCs. iPSC-derived neurons are structurally and functionally mature, and capable of forming electrophysiologically active synaptic networks. Using additional transcription factors during the induction process, it has also been possible to direct differentiation into specific neuronal subtypes, such as dopaminergic neurons. As iPSCs are a relatively new technology, preclinical animal model transplantation studies are few. One study in an ischaemic stroke rodent model demonstrated that human iPSC-derived NSCs were able to improve neurological function and reduce pro-inflammatory factors through a neurotrophin associated bystander effect. In another recent study, following intra-hippocampal transplantation into a transgenic AD mouse model, human iPSC-derived cholinergic neuronal precursors survived, differentiated into phenotypically mature cholinergic neurons, and reversed spatial memory impairment. iPSCs technology allows for the production of autologous pluripotent stem cells, thereby avoiding both the ethical limitations and immune rejection issues of nonpatient-specific sources. Long-term survival and efficacy of autologous iPSC-derived dopaminergic neuronal transplantation has been demonstrated in a simian Parkinson's disease model, with improved motor activity and function, and extensive cell survival and engraftment at 2-year post-operation. Using iPSC-derived neurons to recapitulate AD pathology *in vitro* has significant applications in the study of pathogenesis and screening for potential therapeutic drugs. As such, they are now the subject of extensive study *in vitro*.

2 Stem Cell Therapy in Heart Diseases

Heart diseases are serious and global public health concern. In spite of remarkable therapeutic developments, the prediction of patients with heart failure (HF) is weak, and present therapeutic attitudes do not report the fundamental problem of the cardiac tissue loss. Innovative therapies are required to reduce mortality and limit or abolish the necessity for cardiac transplantation. Stem cell-based therapies applied to the treatment of heart disease is according to the understanding that natural self-renewing procedures are inherent to the myocardium, nonetheless may not be adequate to recover the infarcted heart muscle. Following the first account of cell therapy in heart diseases, examination has kept up to rapidity; besides, several animals and human clinical trials have been conducted to preserve the capacity of numerous stem cell population in advance cardiac function and decrease infarct size.

Throughout the past 15 years, various preclinical and clinical studies investigated the capability of numerous stem cells to progress cardiac function and weaken adversative left ventricular (LV) remodeling in heart diseases. In spite of this quick development, various essential questions arise which need to be determined, beside the fact that up to now no operative cell therapy in patients with cardiomyopathy has been reported. After two decades of focused investigation and efforts of clinical trials, stem cell-based therapies for cardiac diseases are not receiving nearer to clinical accomplishment[4].

2.1 Stem Cell-Based Therapies

In the previous years, excessive efforts have been conducted to induce the least aggression and more efficient sources of human cardiomyocytes for different application, particularly for myocardial regeneration. The pluripotent stem cell seems to be an appropriate candidate, due to proliferate properties and ability to differentiate various cell types including cardiomyocytes. Stem cell based therapy has the probability to activate endogenous regenerative processes, containing the recruitment of resident stem and progenitor cells and the motivation of cardiomyocyte proliferation. Secretion of soluble factors is the predominant mechanism of stem cell mediated heart regeneration. Cytokines and growth factors like transforming growth factor (TGF)-β, stromal cell-derived factor (SDF)-1, and vascular endothelial growth factor (VEGF), can be secreted by transplanted stem and progenitor cells into the intestinal space or bloodstream that stimulates numerous regenerative processes, for instance, neovascularization, activation of tissue intrinsic progenitor cells, decreased apoptosis of endogenous cardiomyocytes, and enrolment of cells of assistance for tissue repair. Some biomarkers, like IL-15, IL-5 and SCF, associate with an improved cardiac function via stem cell treatment representing that higher levels of specific circulating cytokines are appropriate as choice principles for cell-based therapies.

The preclinical trials of stem cell therapy in HF as well as the advantages and disadvantages of these cells for cardiac regeneration are summarized in Tab. 3 – 2 and Tab. 3 – 3, respectively.

Tab. 3 – 2 Preclinical Trials of Stem Cell Therapy in HF

Cell type	Animal model	Study design	Outcome
iPSCs	Porcine model	Iron oxide-labeled hiPSC-CMs cell sheets combined with an omentum flap; Epicardial via the median sternotomy	Increase vascular density in the transplanted area; Increase cardiac troponin T-positive cells
	Porcine model of acute MI	hiPSC-derived cardiomyocytes, endothelial cells, and smooth muscle cells, in combination with a 3D fibrin patch loaded with IGF encapsulated microspheres	Increase the cardiac function
	Mice model of acute MI	iPSC-derived EVs	Improve LV function
	—	Intramyocardial	Improve vascularization; Ameliorate of apoptosis and hypertrophy
ESCs	Sheep with MI	Cardiac-committed mouse ESC	Improve LVEF
CDCs	Mini-pig	Intracoronary	Decrease Infarct size; Improve hemodynamic parameters; Decrease LV remodeling
	Murine model of reperfused MI	CDCs; Intracoronary	Lessen of adverse LV remodeling and dysfunction; Improve global LV systolic and diastolic function; Decrease LV dilation and LV expansion index
	Rat	CPCs were labeled with EGFP; Intracoronary	Lessen fibrosis in the noninfarcted region; Improve LV function; Increase expression of cardiac proteins by endogenous CPCs
	Pigs with post infarct LV dysfunction	cardiosphere-derived cells; Intracoronary	Increase LVEF; Decrease LV remodeling; Improve regional and global LV function; Promote cardiac and vascular regeneration

CDCs: cardiac stem cells; CPCs: cardiac progenitor cells; EGFP: enhanced green fluorescent protein; EVs: extracellular vesicles; FAC: fractional area change; hiPSC-CMs: human induced pluripotent stem cell derived cardiomyocyte; hiPSC: human induced pluripotent stem cell. LV: left ventricular; LVEF: left-ventricular ejection fraction; MI: myocardial infarction.

Tab. 3 – 3 Characteristics of different types of stem cells used for cardiac regeneration

Cell type	Advantages	Disadvantages
iPSCs	Pluripotent differentiation and self-renewal; Low ethical concerns; Easily accessible source tissue; Robust myocardiogenic capacity	Teratoma formation; Immunologic rejection; Limited genome editing technology; Possible genomic instability; Untested in clinical setting
ESCs	Pluripotent differentiation and self-renewal; Easily generation of cell lines; Incorporate electromagnetically into the host myocardium	Ethical issues; Allogenic only; Teratoma formation; Genomic instability; Lack of availability; Immunologic rejection
CDCs	Autologous transplantation; Multipotent; Safety in clinical trials; Low risk of tumorigenicity; Short culture period (weeks) is required to produce CM	Restricted cell quantity; Access from invasive myocardial biopsies; Inadequate cell characterization; Stem cell pool appears to undergo senescence
Skeletal myoblasts	Easy to obtain from muscle biopsies; Autologous transplantation; Resistance to ischemia; Low ethical concerns; Low risk of tumorigenicity	Lack of functional cardiomyocytes differentiation; Risk of ventricular arrhythmias; Low long-term survival rate; Invasive isolation procedure

CM: cardiomyocyte.

2.1.1 Induced Pluripotent Stem Cells

iPSCs are pluripotent stem cells produced from adult somatic cells over a genetic reprogramming process. iPSCs have properties similar to ESCs which is the capacity of self-renewal and differentiation potentiality into several types of cell fates like cardiac myocytes. The ability of iPSCs to maintain patient-particular genomic, transcriptomic, proteomic, metabolomics, and additional personalized information, makes it vulnerable in the field of disease modeling and treatment approaches based on personalized medicine. Human induced pluripotent stem cells (hiPSCs) suggest an exceptional prospect to study human physiology and disease at the cellular level[5].

Production of particular consensus molecular subtypes (CMSs) from hiPSCs will likewise permit more detailed identification of diseases that differently influence both the ventricles (e. g. hypertrophic cardiomyopathy), dilated cardiomyopathy, arrhythmogenic right ventricular cardiomyopathy, or atria (e. g. atrial arrhythmias). Through merging the powers of elementary scientific investigation, biomedical documents knowledge, genome editing, and transformative biomedical stages such as hiPSCs, we could essentially initiate to comprehend personal difference in disease advance, progress, and reaction to the particular treatment.

Functional cardiomyocytes have been produced from both mouse iPSCs and hiPSCs, whereas ultimate differentiation of iPSCs to completely matured cardiomyocytes *in vitro* is still an unsuccessful aim. The main concern about hiPSC is the immature state of discriminated derivatives, and numerous signals are under examination to permit these cells to mature into an adult human heart[6].

The transcription factors of c-Myc, Oct4, and Klf4 make these cells be identified as oncogenes that can create teratoma. Innovative approaches comprising transitory expression of the reprogramming factors, might prevent this obstacle, however, the pluripotent feature of these cells can stimulate tumorigenesis. Another complication involved is the ability of iPSC generation, since the variableness exists among every cell lines. By the uprising in the development of equipment in this ground, it is probable that these methodological impediments will get inspired, and iPSC-based practices will demonstrate to be beneficial for the treatment of heart diseases, while iPSCs are not yet carried for clinical use. After that, major genomic instabilities in iPSC lines containing epigenetic memory, unusual methylation patterns and mutations have been described due to variants in parenteral somatic cells or happening through the reprogramming procedure and culturing time[7].

A new finding indicates the restoration effect of small membrane enclosed droplets that transport biologically active molecules, and genetic materials including stem cell-specific RNAs and proteins from parent cells; called extracellular vesicles (EVs). Developing confirmation recommends that stem cell-derived EVs such as iPSC-derived EVs permit to increase and moderate endogenous protection through transportation of the cargo to numerous cardiovascular cells and motivate reparation of the procedure cells. iPSC-derived EVs contain various population of non-coding RNAs including miRNAs and proteins which could impact the cell survival and proliferation. In a study, mice have been used and were injected with only iPSCs, exhibited teratomas, while on the contrary, none of the mice injected with iPSC-EVs has developed into teratomas.

2.1.2 Embryonic Stem Cells

ESCs are pluripotent cells collected from the internal cell bulk of human blastocysts, the inner cell mass (ICM) in the preimplantation embryos. ESCs have the ability to discriminate into cells of all 3 germ layers, that is, ectoderm, endoderm, and mesoderm when cultivated as 3-dimensional cystic aggregates (embryoid bodies). Cardiomyocytes are known to be originated from the mesoderm layer. Stage specific embryonic antigen-1 (SSEA-1) and SSEA-4, TRA-1-60 and TRA1-81 antigens, Frizzled proteins (Fzd 1 – 10), teratocarcinoma-derived growth factor 1 (TDGF-1) proteins are expressed in human ESCs[7].

Numerous procedures have been effectively advanced to stimulate produce cardiomyocytes from ESCs *in vitro*. Even though the generation of completely matured cardiomyocytes in enormous produces and with great purity is yet impracticable, these studies confirmed influential cardiogenic probable of ESCs. Nevertheless, clinical trials of these cells have been vulnerable by major problems, comprising ethical issues, genetic variability, the risk of immune rejection and tumorigenic possibility. Actually, several first studies theorize that the cardiac environment is adequate to stimulate the differentiation of ESCs into cardiomyocytes. Though, this recommendation has been disproven, from the time when the creation of teratomas was discovered subsequently intramyocardial injection of undifferentiated ESCs. Human ESC-derived cardiomyocytes display adult cardiomyocytes morphology expressing sarcomeric proteins. The possibility of derivation of cardiomyocytes from ESCs has enlightened the interest in exploration of their impacts in cardiomyopathy.

Cardiomyocytes markers are categorized for two phases of early and late differentiation. After 5 – 6 days of differentiation, the markers of GATA binding protein 4 (GATA-4), insulin gene enhancer protein (ISL1), and kinase insert domain receptor (KDR) become highly expressed. Later at day 8 – 9, expression level of markers NK2 homeobox 5 (Nkx 2.5), T-Box 5 (TBX5), myocytes-specific enhancer factor 2C (MEF2C), HAND1/2 is in peak, followed by an increase of myofilament genes [troponin T2 (TNNT2) and myosin heavy chain 6 (MYH6)] at day 8 – 10. Preclinical investigations have emphasized the fact by which retroviral-based gene transfer could assist to avoid cell-based therapy by directing of cardiac fibroblast in the infarct scar to produce useful myocytes. Remuscularization in the infarct scar in a mouse model possibly will result in ventricular dysrhythmias. The dependence on retrovirus technology in this method prevents direct clinical transformation; but this procedure may work as a plan to renew the fibrotic myocardium in patients with ischemic cardiomyopathy, and inhibit disease progress. Nonetheless, published documents propose the possible usefulness of cardiomyocytes made *in vitro* inside the infarct scar possibly will also be accounted in patients with arrhythmia and tachycardia.

Recently, human ESC-derived differentiated cells have been used in patients with spinal cord injury and ocular diseases. hESCs were directed into cardiac cells using fibroblast growth factor receptor inhibitor (FGFRI) and bone morphogenetic protein 2 (BMP2). Cells reacting to these signals express the cardiac transcription factors including ISL-1 and the SSEA-1. Moreover, hESC-derived cells expressing receptor tyrosine kinase-like orphan receptor 2 (ROR2) +, $CD13^+$, vascular endothelial growth factor receptor 2 (KDR) +, platelet-derived growth factor receptor A (PDGFRα) + have been led to produce endothelial cells, vascular smooth muscle cells and cardiomyocytes *in vitro*. The potentiality of these precursor cells is self-renewal and maintenance the capable of differentiating into cardiovascular lineages *in vitro*. This cell population possibly will provide an innovative origin of cells for cardiac regenerative medication.

2.1.3 Cardiac Stem Cells

The detection of mature cardiac stem cells (CDCs) and their potential to renovate cardiac tissue has been brought to attention in the recent years. The expression of KDR in the pool of c-kit + CDCs differentiates myocytes progenitor cells (KDR-) and vasculogenic progenitor cells (KDR +), both of which can differentiate into cardiomyocytes, endothelial cells, and smooth muscle cells; nevertheless, myocytes progenitor cells have a superior susceptibility to produce cardiomyocytes, while vascular progenitors differently obtain vascular smooth muscle cell and endothelial cell fate. c-kit + CDCs sequestered from the grown-up rat heart were revealed to be self-renewing, multipotent, clonogenic and demonstrated all the features of stem cells. Moreover, after being inoculated into the damaged myocardium these cells, were reported to be capable of reestablishing the cardiac structure and activity. In the heart, a small fraction of c-kit + CDCs expresses the transcription factors Nkx2.5 and GATA-4, representing their commitment to the myogenic lineage. Significant mechanism for excretion of cytokines and growth factors by CDCs that could apply paracrine activities on endogenous CDCs, lead them to proliferate and discriminate into adult cardiac cells. c-kit + CDCs considerably improved angiogenesis post MI in a paracrine manner by the secretion of VEGF[7].

2.1.4 Skeletal Myoblasts

Skeletal muscle degenerative is observed in a variety of chronic diseases comprising chronic HF. Muscle wasting is present in about 70% of chronic HF patients. A number of theories have been put forth to describe the HF-related skeletal muscle losing, some of which are physiologic, containing prolonged immobilization and malnutrition, or pathologic, such as insulin resistance, damaged myogenesis and inflammation. Skeletal myoblasts are derivative from skeletal muscle progenitor cells (satellite cells) with regenerative capacity. Following a damage in the muscle, these progenitor cells experience multiplying and stimulate renewal by discriminating into myotubes and fresh muscle fibers. The first effective transplant of skeletal myoblasts into the damaged human heart was experimented in 1994. For that, skeletal myoblasts differentiated into myotubes and then inter-connected with decrease of adversarial ventricular renovation, reduction of interstitial fibrosis, and increase of cardiac act. After the process of differentiation, the skeletal myotubes miss the capability to form gap junctions because of their deficiency in expressing main gap junction proteins such as connexin-43 and N-cadherin, which results in the lack of electrical integration with the myocardium and enhanced risk of ventricular arrhythmia[8].

2.2 Brief Summary

Epidemic abundance of heart disease carries on rising over the time. Many patients with injured myocardium have to suffer symptoms of HF which could be lessened by medicines, repeated coronary revascularization and cardiac transplantation. A number of the patients fail to satisfactorily respond to these therapies; besides, many patients have no chance of getting cardiac transplantation, because of very restricted accessibility of heart donors and other complications involved. The revolt in stem cell technology, together with the improved considerate of the endogenous process essential organ repair, has delivered the scientific basis for the progress of regenerative attitudes. New therapies such as stem cell therapies for cardiac regeneration have been progressively examined in the recent years.

Considering the present position of cell-based therapies for heart disease, it is significant to retain a historic perception. Many significant concerns (e. g. mechanisms of stem cells action, long-standing engraftment, optimum cell types, dosage, route, and rate of recurrence of cell administration) keep on being determined, and no cell therapy has been determinedly revealed to be operative. It is factual that the detailed mechanism of stem cells' actions remain uncertain and their efficiency in heart diseases has not been recognized. Currently, it is not pure whether and which type of stem cell or technique of cell delivery is better than others. It is expected that increasing amount of comparative studies will be the emphasis issue of impending basic and clinical studies in cardiovascular regenerative medicine.

3 Stem Cell Therapy in Diabetes Mellitus

Diabetes mellitus (DM) type 1 and type 2 have become a global epidemic with dramatically increasing incidences. Poorly controlled diabetes is associated with severe life-threatening complications. Beside traditional treatment with insulin and oral anti-diabetic drugs, clinicians try to improve patient's care by cell therapies using ESCs, iPSCs and adult MSCs. ESC displays a virtually unlimited plasticity, including the differentiation into insulin producing β-cells, but they raise ethical concerns and bear, like iPSCs, the risk of tumors. iPSCs may further inherit somatic mutations and remaining somatic transcriptional memory upon incomplete reprogramming, but allow the generation of patient/disease-specific cell lines. MSCs avoid such issues but have not been successfully differentiated into β-cells. Instead, MSCs and their pericyte phenotypes outside the bone marrow have been recognized to secrete numerous immunomodulatory and tissue regenerative factors. MSCs are currently the most investigated cells in DM-related trials while clinical testing of ESCs has just started.

3.1 β-Cell Replacement

Patients with autoimmune T1DM experience a loss of insulin producing pancreatic β-cells and rely on daily insulin injections. Despite modern insulin therapies, exogenous application of insulin can never be as accurate and dynamic like insulin secretion from endogenous β-cells and therefore can only partially reduce the risk for the development of microvascular (i. e. nephropathy, retinopathy) or macrovascular (i. e. coronary artery disease, peripheral artery disease, cerebrovascular disease) complications. Additionally, efforts to develop effective immunosuppressive treatments to prevent β-cell loss before disease onset had limited success so far[9].

Clinical islet transplantation aims to re-establish endogenous insulin secretion and has been steadily refined since its beginning in the 1980s. Compared to standard insulin therapy, islet transplantation more efficiently improved glycemic control and progression of retinopathy, and resolved hypoglycemia even in patients with only partially remaining graft function.

Transplantation of whole pancreases is an established alternative to islets and both procedures display advantages and limitations. The standard procedure of islet infusion into the liver is much safer with less complication than pancreas transplantation which is considered a major surgery with accordingly enhanced risks for the patient. Thus, pancreas transplantation is rarely performed alone and is most commonly combined with kidney transplantation in patients with T1DM and end-stage renal disease. The major obstacle of the less risky islet transplantation is the limited graft survival[10].

Whole pancreas transplantation strictly depends on high quality organs while generation of insulin producing β-cells from SC has the potential to solve the problem of limited availability of donor material for islet transplantation. However, the need for immunosuppression reserves both pancreas and islet transplantation as a therapeutic choice to a limited patient population such as brittle diabetics with life-threatening hypoglycemic events or subjects who anyway receive immunosuppression (e. g. kidney transplantations for renal failure due to diabetic nephropathy).

3.2 Embryonic Stem Cells

The differentiation of hESCs into functional β-cells is not trivial since transforming processes have to mimic complex embryonal organogenesis *in vitro*. Differentiation protocols therefore established a number of factors and inhibitors that modulate molecular pathways in an exact sequential timing to resemble the natural development of pancreatic β-cells. Generally, hESCs were firstly differentiated into definitive endoderm cells and then sequentially into primitive gut tube and posterior foregut, pancreatic endoderm and finally β-cells using multiple specified media and supplementation in each step.

A major step in differentiation of hESCs towards β-cells is the expression of the transcription factors pancreatic and duodenal homeobox 1 (PDX1) and NK6 homeobox 1 (NKX 6.1), which are markers of pancreatic endoderm and endocrine precursor cells. It has been shown that comparable human embryonic pancreas tissue from fetal weeks 6–9, which contained very few β-cells at that stage, is capable to mature into functional β-cells after transplantation into non-obese diabetic mice with severe combined immunodeficiency (NOD/SCID mice).

Based on these findings, *in vivo* maturation of PDX1+/Nkx6.1+ progenitors into β-cells has been recognized to be more efficient than mimicking this months-long process *in vitro*. Furthermore, scalable *in vitro* differentiation of ESCs into endocrine pancreatic precursor cells is to date more robust and less complicated than generation of fully functional β-cell phenotypes by advanced protocols.

Although ESC-derived progenitors are hypoimmunogenic, transplanted cells are challenged by adaptive immune responses such as local inflammation and rejection. In addition, once maturated into insulin producing β-cells, graft cells will be attacked by persistent autoreactive T cells in patients with T1DM. Several studies have demonstrated that macroencapsulation protects embedded cells by isolation from immune responses and thereby avoids rejection and the need for immunosuppression. Furthermore, macroencapsulation prevents escape of embedded cells into the body. This is an important safety issue since any ESC transplanted in an undifferentiated state bears the potential of malignant transformation[11].

3.3 Induced Pluripotent Stem Cells

Following the principle route of pancreatic development as used for ESCs, researchers have successfully differentiated iPSCs into functional β-cell phenotypes and also established scalable production of both endocrine pancreatic progenitors and β-cells. Nevertheless, further refinement of procedures is still an issue and new tools were created for the identification of compounds and conditions which enhance yield and functionality of generated β-cells. For example, hiPSCs expressing the fluorescent reporters Venus and m-Cherry under the control of intrinsic neurogenin 3 and insulin promoters have been generated for screening of differentiation efficiency. These cells have served to identify an inhibitor of fibroblast growth factor receptor 1 (FGFR1) that, while blocking the early development of pancreatic progenitors, promoted the terminal differentiation of pancreatic endocrine progenitors into endocrine cells including β-cells.

However, due to their origin in adult somatic cells, iPSCs can inherit somatic mutations and incomplete reprogramming can maintain somatic transcriptional memory including cancer associated gene activity. These dangers currently do not define them as the first choice for clinical use but, more importantly, iPSCs enable the successful generation of patient-specific cell phenotypes that allow to recapitulate disease processes *in vitro* and can serve as platforms for drug development and testing. For example, researchers successfully generated an iPSC line from a patient carrying a hepatocyte nuclear factor 1α (HNF1-α) mutation resulting in maturity-onset diabetes of the young type 3 (MODY3). In the near future, patient-specific cell lines will help to develop disease-related models that overcome the obstacle of species differences between human subjects and animal models[12].

3.4 Mesenchymal Stem Cells

MSC could be easily expanded *in vitro* without significant loss of their mesenchymal differentiation capacity or their humoral secretion. Moreover, MSCs are immune-privileged because they express very low levels of MHC class I and no MHC class Ⅱ which normally prevents or strongly reduces immune responses. In clinical use since 1995, MSCs are considered clinically safe and both administration of autologous and allogenic MHC-mismatched MSCs is generally well tolerated and clinically effective.

In many patients with T1DM a minor portion of insulin producing β-cells survive but cannot recover unless thereby induced autoimmune responses are blocked. MSCs mediate immune tolerance that aims to enable partial recovery of remaining β-cell mass or to reduce and delay the β-cell destruction during new-onset of T1DM.

In T2DM, the anti-inflammatory features of MSCs were used to ameliorate chronic low-grade inflammation which has been recognized as an important cause of insulin resistance and β-cell dysfunction. These features in combination with secretion of pro-angiogenic factors should improve engraftment and survival of transplanted islets.

The International Society for Cell Therapy (ISCT) defined clinically useful MSCs by mesenchymal differentiation (into bone, cartilage and fat), plastic-adherent growth *in vitro* and expression of CD73, CD90 and CD105 in the absence of hematopoietic surface markers. Notably, expansion of plastic-adherent BM cells favors the expansion of nonclonal stromal cell-enriched populations, often misinterpreted as pure stem cell fractions, which contain varying percentages of true MSCs, e. g. depending on donor age, and thereby plausibly exhibit different clinical effectiveness. This demands a reliable assay, such as the CFU-F assay, and careful evaluation may allow the identification of the percentage of stem cells and their multilineage potential in each batch of nonclonal MSCs. Potentially, efficacy of MSCs populations could be further enhanced by selection via additional markers such as stromal (STRO)-1, CD146, alkaline phosphatase, CD49a, CD271 and VCAM1[13].

3.5 Brief Summary

Differentiation of ESCs and iPSCs has meanwhile reached clinical large scale production and current developments of macroencapsulation may provide clinical safe usage of these cells that demonstrate otherwise potentials for tumor development. Macroencapsulation prevents the escape of embedded cells into the body and subcutaneously transplanted devices could be retrieved and removed easily[14].

MSCs are clinically safe and several trials exist though they are limited in number and investigated patients. Currently MSC-based therapy is no cure but shows a potential to ameliorate DM since most studies report decreased requirement of exogenous insulin and/or anti-diabetic drugs. In this regard MSCs may be best used with diabetic patients that have severe problems in controlling glycemia by conventional therapies; e. g. patients with brittle DM (Tab. 3-4).

Tab. 3 – 4 Summary and comparison of stem cells in cell therapy of diabetes mellitus

Items	ESCs	iPSCs	MSCs
Cell type and origin	Embryonic stem cells; Inner cell mass of the blastocyst	Adult somatic cells; Reprogramming in vitro	Adult stem cells; Endosteal (BM) and perivascular niches (all tissues)
Characteristics	Pluripotent; Generate all germ layers: ectoderm, endoderm, and mesoderm	Pluripotent; Generate all germ layers: ectoderm, endoderm and mesoderm	Multipotent; Generate mesenchymal lineages: bone, cartilage, fat and muscle; Maintain HSC niche and hematopoiesis
Ethical concerns	Use of embryos	No	No
Differentiation into pancreatic β-cells	Yes	Yes	Insulin + cells with limited secretory or proliferative capacity (experimental)
Cell therapeutic options	β-cell replacement	β-cell replacement; Patient-specific cell lines	Secreted factors with immunomodulatory, angiogenic and tissue regenerative properties
Advantages	Large-scale production of pancreatic endoderm, endocrine progenitors and fully functional β-cells	Large-scale production of pancreatic endoderm, endocrine progenitors and fully functional β-cells	Easy isolation and in vitro expansion; Low immunogenicity allows allogenic transplantation without immunosuppression; Clinical safe and well tolerated
Limitations	Tumorigenic if incompletely differentiated	Tumorigenic if incompletely differentiated; Somatic mutations; Incomplete reprogramming maintains somatic transcriptional memory	Current clinical protocols are not standardized and exhibit potential for improvements; Only a small proportion of systemically injected cells engrafts in injured target tissues
Status	First-in-man trial currently investigates clinical safety and efficacy of ESC-derived pancreatic endoderm; Macroencapsulation avoids tumorigenicity by preventing the escape of embedded cells into the body	Currently not safe enough for clinical usage; Patient-specific cell lines allow investigation of disease processes *in vitro* and represent a platform for drug testing	Completed clinical trials collectively report on reduced requirement for exogenous insulin; Greatest benefit for patients with problems in controlling glycemia by conventional therapy; More clinical trials in progress

4 Stem Cell Therapy in Osteoarthritis

Osteoarthritis (OA) is a progressive and up-to-date irreversible disease, representing one of the most prevalent musculoskeletal disorders and the most common rheumatic disease. OA leads to articular cartilage degeneration resulting in pain and stiffness, and in advanced stages may provoke the loss of joint mobility. Although OA is mainly associated with aging, multiple risk factors can contribute to its appearance including sex, obesity, genetics, joint injury/trauma, overuse of the joint, and underlying anatomical and orthopedic disorders, as well as other co-morbidities such as diabetes, metabolic, and endocrine diseases. Although any joint can be affected, it is mainly suffered in the knees, hips, hands, ankles, and neck. The impact of OA in the knee joint is especially problematic for the perfor-mance of the usual daily activities. As life expectancy is expected to increase in the next few coming decades, agerelated musculoskeletal disorders like OA will become a major health problem in societies, constituting one of the most relevant causes of incapacity in the elderly and middle-aged people[15].

In terms of the pathobiology of OA, it has been reported that the main cause of cartilage destruction is an imbalance between catabolism and anabolism. Increased activity of catabolic signaling pathways and extracellular matrix (ECM) degrading enzymes leads to chondrocyte hypertrophic differentiation, matrix degradation, and vascular invasion. From a molecular point of view, the aging process of chondrocytes implies a lower response to growth factors accompanied by a greater expression of pro-inflammatory cytokines such as tumor necrosis factor alpha (TNF-α) and interleukin-1 (IL-1), which entails an increase in the synthesis of matrix metalloproteinases resulting in the loss of the ECM. Moreover, OA could induce alterations in chondrocyte signaling activities resulting in hypertrophic phenotype through an increase in their differentiation process. OA dramatically diminishes the quality of life of the patients and greatly complicates their daily life. Currently, the available conventional treatments provide little option to reverse the progression of cartilage degeneration, giving symptomatic relief toward inflammation and pain. In mild cases, the patients are often treated by medications, diet, exercise, and rehabilitation/physiotherapy, whereas in advanced or end-stage OA total joint arthroplasty is the ultimate solution proposed. However, this prosthetic replacement could potentially give rise to infection, which constitutes one of the most serious of complications of the procedure. Other surgical techniques that can be performed in order to repair and stimulate the renovation of articular cartilage and therefore prevent progression toward OA include arthroscopic lavage and debridement, marrow stimulation by microfracture, abrasion or drilling of the subchondral bone plate, osteochondral auto/allografting, and autologous chondrocyte implantation (ACI). Although surgical methods have been proposed to restore normal joint function and minimize further degeneration, they all are associated with complications, side effects, and unsatisfactory progress, leaving in many cases, the lesions inadequately treated and often not offering a long-term clinical solution. Other disadvantages they present are the high costs associated with some of these procedures and their invasive nature.

Cell-based therapy has emerged as a promising approach toward cartilage regeneration. ACI was the first cell therapy applied and it has been frequently used for the restoration of cartilage defects, demonstrating excellent short to mid-term repair.

Nevertheless, this approach presents some drawbacks related to chondrocyte dedifferentiation and the fact that chondrocytes fail to maintain their chondrogenic phenotype during the expansion phase. In a worst-case scenario, these issues might result in fibrocartilage formation instead of hyaline cartilage, which is not fully functional (i. e. it has weaker biomechanical properties) nor long lasting and also has higher permeability, therefore, can not withstand the demands of everyday activities. One of the main reasons for cell dedifferentiation is the different microenvironments created between the two-dimensional expansion *in vitro* and the real *in vivo* three-dimensional (3D) microenvironment.

Besides, the hypertrophic differentiation of some chondrocytes could lead to endochondral ossification forming new bone instead of cartilage in articular cartilage defect site via matrix calcification and vascular invasion.

This technique has further been developed into the use of scaffolds as supports for cell growth, e. g. matrix-induced autologous chondrocytes implantation (MACI). However, both ACI and MACI present another inherent disadvantage, as they require two surgical interventions (one to remove healthy cells from the donor site and the other to implant the cells/matrix), that may cause further cartilage damage and degeneration, as well as several limitations concerning patient mobility and life quality. Moreover, its application has so far been limited to focal cartilage defects caused by injury while generalized cartilage loss seen in OA has not been addressed by this technique yet. These limitations and disadvantages had fueled progress and research with other cell sources for cell-based therapies in order to achieve a permanent restoration of articular cartilage.

4.1 Stem Cell Therapy

MSCs have the ability of interacting with numerous types of immune cells regulating their functions in two different ways. First, MSCs can interact with the cells of the immune system directly by cell-to-cell contact thanks to cellular surface molecules. Second, they can exert their immunomodulatory properties via releasing soluble molecules (secretome) such as cytokines, growth factors, immunomodulatory factors, or extracellular vesicles (EVs) to the extracellular environment. The pro-inflammatory environment that facilitates OA progression within otherwise healthy joints could be attenuated through the immunomodulatory properties of MSCs.

MSCs can also promote protection against apoptosis induced by trauma, oxidative stress, and other types of injuries through the secretion of various growth factors such as IGF-1, TGF-β1, stanniocalcin-1, IL-6, the antioxidative molecules erythropoietin and heme oxygenase (OH) -1. They also present anti-fibrotic properties probably exerted through the secretion of HGF, adrenomedullin, and bFGF.

MSCs also release EVs that could become mediators of the paracrine action in regenerative medicine. Some of these EVs display important immunosuppressive role in T cell and B-lymphocytes populations, and can also inhibit the inflammatory functions of monocytes and macrophages. Exosomes derived from MSCs are enriched in miRNAs and mRNAs. In a model of kidney injury, most of the miRNAs derived from Wharton's jelly derived MSCs have the capacity of inhibiting the expression of CX3CL1, a potent macrophage chemoattractant located in the endothelial cell surface. In this way, they suppress the accumulation of pro-inflammatory macrophages in the kidney. An interesting way of communication among T cells and MSCs is carried out by developing tunneling nanotubes of communication derived from the T cells that would pass this type of vesicles. Moreover, the EVs can also exert tissue regenerative activity by being highly pro-angiogenic, antiapoptotic, and anti-fibrotic.

Recently, it has been proposed that given the evidences of the therapeutic properties of EVs, in the treatment of OA, the use of EVs could suppress the function of the MSCs by themselves. In an osteoarthritic rat model, synovial MSCs were modified to overexpress miR-140-5p, which regulates the cartilage homeostasis and development. The improved EVs led to a significant inhibition of cartilage regeneration.

Another cell source is adult chondrocytes isolated from articular cartilage, nasal septum, and costal cartilage. Though interesting, their use is time-consuming as these cell phenotypes require prolonged expansion periods until reaching the sufficient number of cells for the transplant, and the prolonged expansion can even sometimes transform chondrocytes into fibroblasts.

4.2 Route of Administration

Scaffold free (injection), one of the most relevant aspects for the success of a cell-based therapy for the treatment of OA, is to control the function and survival of the injected cells in the following days and weeks after the treatment. In most cases, MSCs are applied directly by intra-articular injection without any carrier both in animal models and in clinical trials. Direct injection offers great advantages for clinical translation owing to its simplicity and ease of application, which provides the possibility of better treatments avoiding at the same time the surgeries and consequently the side effects associated to them, especially for elderly with comorbidities. However, direct delivery can lead to limited stem cell engraftment at the injured site due to a leakage of the cell suspension during injection and a probably reduced viability as a consequence of the cell damage suffered during the procedure.

This damage can be caused by the shock experienced by the cells when they are removed from culture media and injected into the host, or due to the damaging local microenvironment in the targeted tissue. In many cases, less than 5% of the transplanted cells remain at the application site within days after cell injection. Consequently, improvement of cell engraftment and viability seems a critical point to take into account to improve the success rate of cell therapies. In order to address this and improve the aforementioned cell viability it has been suggested to embed SCs within protective carriers (scaffolds and hydrogels), which not only may protect the cells and help to retain them in the proper place but also may help to provide a suitable microenvironment for supporting cell viability, function and phenotype, and facilitating optimal cartilage regeneration and hyaline cartilage formation.

5 Stem Cell Therapy in Wound Healing

Wound healing is an intricate process comprised of distinct but overlapping stages. Aberrations in any of the stages of healing can lead to delayed wound closure and pose a significant burden on patients, their families, and the healthcare system. Certain patient groups, such as diabetic patients and the elderly, are more susceptible to the development of chronic wounds. Several treatment options for wound care currently exist, but few can efficiently reverse the deficiencies that contribute to chronic wounds and restore the tissue to its pre-injured state. Patient groups at risk of developing chronic wounds are rapidly growing, and the need to develop more effective wound care options is urgent[16].

5.1 Stages of Wound

Cutaneous wound healing is an intricate process that requires the coordination of a number of processes. Briefly, there are four stages of healing: hemostasis, inflammation, proliferation, and remodeling. During hemostasis, bleeding is stopped through vasoconstriction and formation of a fibrin clot that serves as a scaffold for incoming immune cells during the inflammatory stage. Pathogens are cleared from the wound by an influx of neutrophils and macrophages. In the proliferative stage, resident fibroblasts migrate into the fibrin clot and proliferate to form contractile granulation tissue and produce collagen. The proliferating fibroblasts deposit extracellular matrix (ECM), and a subset of fibroblasts differentiates into myofibroblasts which draw the wound margins together. Endothelial cells concurrently proliferate, migrate, and reorganize with structural support from pericytes to form new blood vessels. Re-epithelialization then occurs, during which epidermal stem cells proliferate to re-establish the epidermal layer. Finally, epidermal appendages are formed through the proliferation of stem cells for sebaceous glands, sweat glands, and hair follicles.

5.2 Chronic Wounds

Wound healing resolves on its own in the majority of cases. In susceptible populations, aberrant healing can lead to substantially delayed healing or non-closure of wounds, causing significant morbidity and mortality. Chronic wounds are defined as those that have failed to proceed through an orderly and timely reparative process to produce anatomic and functional integrity of the injured site and are classified into four categories: pressure ulcers (PUs), diabetic ulcers, venous ulcers, and arterial insufficiency ulcers. Compared to normally healing wounds, chronic wounds exhibit increased levels of proinflammatory cytokines, proteases, reactive oxygen species (ROS), and senescent cells, and a depleted population of stem cells. After injury, microorganisms and platelet-derived growth factors such as TGF-β or ECM molecules recruit immune cells, which amplify the release of proinflammatory cytokines that persist in the chronic wound.

Immune cells in the wound produce ROS, which further damages ECM proteins, activating additional proteases and inflammatory cytokines. Protease levels in chronic wounds exceed levels of their inhibitors, leading to their dysregulation and degradation of ECM, growth factors, and their receptors. This dysregulation stalls the wound in the inflammatory phase and prevents it from entering the proliferative phase.

Application of cellular tissue has also been shown to improve wound healing outcomes. Autologous skin grafts are used in burns and following extensive wounding to quickly replace large surface areas of tissue. Current efforts are focused on bioengineering these skin substitutes but continue to be very costly. Commercially available cellular tissue is typically comprised of both differentiated and undifferentiated cellular populations. Most often cellular tissue includes keratinocytes and fibroblasts, which promote deposition of ECM components and release growth factors to stimulate recruitment of endogenous cells, promote neovascularization, granulation tissue formation, and re-epithelialization. Currently, Grafix, comprised of cryopreserved amniotic membrane, is the only FDA approved wound healing product containing MSCs and has to date exhibited the greatest efficacy of any cell-based wound healing product, and is indicated for many types of acute and chronic wounds, including DFUs, VLUs, PUs, burns and epidermolysis bullosa.

5.3 Types of Stem Cells

An array of modalities are available for the treatment of chronic wounds, but limited efficacy of current treatments persists. There has been considerable interest in using stem cell therapies to promote wound healing. Stem cells are defined by their ability to self-renew and their multipotency, or ability to differentiate into more than one cell type. They have been shown to secrete cytokines and growth factors involved in immunomodulation and regeneration, making them a promising agent for correcting the underlying biological deficiencies observed in chronic wounds. Stem cell types ranging from immature pluripotent stem cells to more restricted multipotent progenitor cells have been explored for their ability to promote wound healing in several animal models and clinical trials.

5.3.1 Embryonic Stem Cells and Induced Pluripotent Stem Cells

ESCs are pluripotent cells that can contribute to all three germ layers. ESCs are isolated from the inner cell mass of the blastocyst, readily expandable in culture, and can be guided with growth factors in supplemented media to differentiate into various cell types, including mature cells involved in wound repair, such as HSCs, differentiated immune cells, and epidermal cells. ESCs have been successfully differentiated into functional keratinocytes to form an epidermal sheet *in vitro* that could be used as a temporary skin substitute in patients awaiting autologous grafts. However, ethical considerations and legal restrictions have limited the clinical use of ESCs. iPSCs are adult somatic cells reprogrammed into pluripotent cells and offer a promising clinical alternative to ESCs, bypassing substantial ethical issues surrounding ESCs use. Adult fibroblasts were first reprogrammed to a pluripotent state by a major advantage of iPSCs is the retention of the genotype of the donor following differentiation, which reduces risks for immune rejection in autologous therapy, and provides opportunities to recapitulate diseases *in vitro* or correct genetic defects to generate healthy autologous tissue.

5.3.2 Mesenchymal Stromal Cells

MSCs are multipotent stem cells that can be derived from various tissue sources, and their therapeutic benefit has been described in several disorders. Their ability to enhance healing is due to their pro-regenerative and immunomodulatory properties through paracrine signaling, as well as their ability to traffic to sites of injury. Importantly, application of MSCs does not elicit an immune response by the recipient. MSCs are characterized by the positive expression of surface markers CD105, CD73, and CD90, and negative expression of CD45, CD34, CD14 or CD11b, CD79a or CD19, and HLA-DR, as established by the International Society for Cellular Therapy (ISCT).

MSCs can inhibit the expression of matrix metalloproteinase (MMP)-1, which prevents the degradation of ECM and may contribute to fibroblast proliferation, a process that is dysregulated in chronic wounds. Finally, there is increasing evidence that secretion of MSC exosomes produces a therapeutic effect.

A significant barrier to clinical use of BM-MSCs is their isolation, which requires obtaining bone marrow aspirates, a procedure that can cause significant patient morbidity and yield an insufficient number of cells required for treatment. An alternative to BM-MSCs are adipose-derived stromal cells (ASCs), which are isolated from lipoaspirates that are frequently disposed of as medical waste following procedures. ASCs isolation from adipose tissue yields a high number of cells and allows clinical application without *in vitro* expansion. ASCs share many of the therapeutic characteristics exhibited by BM-MSCs, and their application in preclinical models and clinical trials for wound healing has produced similar successes, resulting in accelerated healing and increased neovascularization of the wound. Human umbilical cord blood (UCB) and extra-fetal tissue, including amniotic fluid, Wharton's jelly, placental tissue and UC lining tissue have also recently emerged as MSCs tissue sources with significant therapeutic potential.

5.4 Brief Summary

Although a number of modalities have been explored, currently available treatments for wound healing are only moderately effective and fail to reliably heal chronic wounds. Thus, there is an urgent need to develop better therapies. Stem cell therapy presents a promising new approach to wound healing, with the potential to regenerate tissue to its pre-injured state. MSCs, which can be easily isolated from a variety of tissue sources, including bone marrow, fat and extra-fetal tissue, are the primary focus of emerging cell therapies due to their therapeutic paracrine activity and potential to home to sites of injury. Several clinical trials are investigating the safety and efficacy of stem cell therapies for the treatment of burns and chronic wounds, but additional basic and clinical research will need to be performed before stem cell therapy can become a mainstream treatment option.

Supplement

List of Abbreviations	
AD	alzheimer's disease
ASCs	adipose-derived stromal cells
BMP	bone morphogenetic protein
CDCs	cardiac stem cells
CM	cardiomyocyte
CPCs	cardiac progenitor cells
DM	diabetes mellitus
EGFP	enhanced green fluorescent protein
EVs	extracellular vesicles

(To be continued)

List of Abbreviations	
FAC	fractional area change
HSCs	hematopoietic stem cells
ICM	inner cell mass
LV	left ventricular
NSCs	neural stem cells
OA	osteoarthritis
SGZ	subgranular zone
SVF	stromal vascular fraction
SVZ	subventricular zone
T1DM	autoimmune DM type 1
UCB	umbilical cord blood

Key Words List	
超磷酸化	hyperphosphorylated
孤雌生殖的	parthenogenetic
低免疫原性	hypoimmunogenic
分泌组	secretome
宏封装	macroencapsulation
化学吸引剂	chemoattractant
回旋	gyrus
肌成纤维细胞	myofibroblasts
肌管	myotubes
肌肉再生	remuscularization
抗凋亡	antiapoptotic
脑室下	subventricular
内嗅	entorhinal
鞘内	intrathecal
趋化因子	chemokines
缺血的	ischaemic
软骨下	subchondral

(To be continued)

Key Words List	
神经营养	neurotrophic
肾上腺髓质素	adrenomodulin
同种异体	allogenic
脱氨酶	deaminase
心肌细胞	cardiomyocytes
心律失常	arrhythmogenic
伊立替康	irinotecan
乙酰转移酶	acetyltransferase
运动	locomotor
造血的	hematopoietic
周细胞	pericyte

References

[1] ZAKRZEWSKI W, DOBRZYŃSKI M, SZYMONOWICZ M, et al. Stem cells: past, present, and future [J]. Stem cell research & therapy, 2019, 10: 68.

[2] SIMARA P, MOTL J A, KAUFMAN D S. Pluripotent stem cells and gene therapy [J]. Translational research, 2013, 161: 284 – 292.

[3] LIN W, HUANG L, LI Y, et al. Mesenchymal stem cells and cancer: clinical challenges and opportunities [J]. Biomed research international, 2019, 2019: 2820853.

[4] KADOTA S, SHIBA Y. Pluripotent stem cell-derived cardiomyocyte transplantation for heart disease treatment [J]. Current cardiology reports, 2019, 21: 73.

[5] MÜLLER P, LEMCKE H, DAVID R. Stem cell therapy in heart diseases—cell types, mechanisms and improvement strategies [J]. Cellular physiology and biochemistry, 2018, 48: 2607 – 2655.

[6] CHO G S, FERNANDEZ L, KWON C. Regenerative medicine for the heart: perspectives on stem-cell therapy [J]. Antioxidants & redox signaling, 2014, 21: 2018 – 2031.

[7] GORADEL N H, HOUR F G, NEGAHDARI B, et al. Stem cell therapy: a new therapeutic option for cardiovascular diseases [J]. Journal of cellular biochemistry, 2018, 119: 95 – 104.

[8] MARDANPOUR P, NAYERNIA K, KHODAYARI S, et al. Application of stem cell technologies to regenerate injured myocardium and improve cardiac function [J]. Cellular physiology and biochemistry, 2019, 53: 101 – 120.

[9] WU H, MAHATO R I. Mesenchymal stem cell-based therapy for type 1 diabetes [J]. Discovery medicine, 2014, 17: 139 - 143.

[10] PELLEGRINI S, PIEMONTI L, SORDI V. Pluripotent stem cell replacement approaches to treat type 1 diabetes [J]. Current opinion in pharmacology, 2018, 43: 20 - 26.

[11] MAHDIPOUR E, SALMASI Z, SABETI N. Potential of stem cell-derived exosomes to regenerate β islets through Pdx-1 dependent mechanism in a rat model of type 1 diabetes [J]. Journal of cellular physiology, 2019, 234: 20310 - 20321.

[12] PAN G, MU Y, HOU L, et al. Examining the therapeutic potential of various stem cell sources for differentiation into insulin-producing cells to treat diabetes [J]. Annales d'endocrinologie (Paris), 2019, 80: 47 - 53.

[13] BOHÁČOVÁ P, HOLÁŇ V. Mesenchymal stem cells and type 1 diabetes treatment. Mezenchymální kmenové buňky a léčba diabetu 1. typu [J]. Vnitrni lekarstvi, 2018, 64: 725 - 728.

[14] DUFFY C, PRUGUE C, GLEW R, et al. Feasibility of induced pluripotent stem cell therapies for treatment of type 1 diabetes [J]. Tissue engineering part B-reviews, 2018, 24: 482-492.

[15] AJANI J A, SONG S, HOCHSTER H S, et al. Cancer stem cells: the promise and the potential [J]. Seminars in oncology, 2015, 42 Suppl 1: S3 - S17.

[16] JOHNSON J Z, HOCKEMEYER D. Human stem cell-based disease modeling: prospects and challenges [J]. Current opinion in cell biology, 2015, 37: 84 - 90.

Chapter 4　Stem Cell/Stem Cell Niche and Tissue Regeneration

第四章　干细胞/干细胞微环境与组织再生

［中文导读］

1978 年，科学家 Schofield 首次提出了干细胞微环境（stem cell niche，stem cell microenvironment）的概念。干细胞微环境的概念是由 20 世纪 70 年代的科学家针对人类造血干细胞（hematopoietic stem cells，HSCs）的特殊微环境而提出来的，他们随后提出了关于表皮、消化道上皮、神经系统和性腺等组织的相似概念。

干细胞微环境，又称干细胞壁龛、干细胞龛，是干细胞的集中存储部位，由干细胞、周围的细胞、细胞外基质及外部信号分子构成。在人体中，干细胞微环境为干细胞的自我更新提供了合适的外部条件，并且抑制干细胞向成熟细胞分化。但是，当组织发生损伤，周围的微环境会传递信号给成体干细胞，引起成体干细胞的自我更新，同时分化为成熟的细胞，对损伤的组织进行修复，促进伤口愈合。干细胞的增殖分化行为一方面被细胞本身预先程序化，另一方面受其所处的微环境调控。干细胞微环境对干细胞的调控是通过多种方式进行的，比如干细胞间的相互作用、干细胞与邻近分化细胞的相互作用、干细胞和黏附因子的相互作用、细胞外基质、细胞因子、生长因子、pH、离子强度等。干细胞微环境对干细胞的调控是一个复杂的网络，具体的调控机制还需要更多的研究。此外，科学家认为干细胞微环境通过调控使干细胞处于静止期并维持分化与自我更新之间的平衡从而避免肿瘤的发生，任何可导致干细胞脱离其壁龛的突变都可能诱发肿瘤。

再生医学在疾病预防与治疗方面具有广阔前景，能解决目前临床上棘手的问题。许多研究人员采用移植干细胞的方法来修复受损组织。此外，随着对干细胞微环境的深入研究，科学家们想通过控制干细胞微环境促进内源性干细胞修复损伤组织。干细胞微环境是高度动态变化的，因此，可以通过小分子物质、生物分子或者生物材料靶向微环境中的特定反应如细胞间相互作用、细胞与细胞外基质作用，来刺激干细胞的增殖和分化，又或者使已经分化的细胞逆分化为干细胞。然而，这种通过靶向干细胞微环境来治疗疾病的方法，目前仍然存在很大的挑战。

1　Introduction to Stem Cell Niche

Stem cells, which are capable of both reproducing themselves (self-renewing) and generating the differentiated cell types that are needed to carry out specialized functions, are responsible for the growth, homeostasis and repair of many tissues. Stem cell behavior, in particular the balance between self-renewal and differentiation, is ultimately controlled by the integration of intrinsic factors with extrinsic factors supplied by the surrounding microenvironment, often referred to as the "stem cell niche".

Schofield put forward the stem cell niche hypothesis in 1978. He proposed that stem cells reside within fixed compartments, or niches, which are conducive to the maintenance of definitive stem cell properties. Thus, the niche represents a defined anatomical compartment that provides signals to stem cells in the form of secreted and cell surface molecules to control the rate of stem cell proliferation, determine the fate of stem cell daughters, and protect stem cells from exhaustion or death. The stem cells are regulated by both intrinsic cues and extrinsic cues from stem cell niche. The identification and characterization of niches within tissues has revealed an intriguing conservation of many components, although the mechanisms that regulate how niches are established, maintained and modified to support specific tissue, stem cell functions are just beginning to be uncovered.

1.1 Definition

Stem cell niche is a specific location in a tissue which harbors stem cells and provides external control of stem cells. The phrase technically refers to the specific part of each general area of tissue which is believed to contain stem cells. Many different locations within the adult human body contain a stem cell niche, such as the bone marrow, skin, brain, liver, and heart. In addition, stem cells exist in abundance in embryos. Ordinarily, stem cells occupy a stem cell niche until they are needed by the body to replace some damaged tissue. Within the niche, the stem cells ordinarily don't divide, remaining quiescent until they are needed. The niche is important to control stem cell function such as proliferation and differentiation and for regeneration, homeostasis and aging.

1.1.1 The Location of the Intestinal Stem Cell Niche

In the human body, the epithelium of the small intestine has a higher self-renewal rate than any other tissue, with a turnover time of less than 5 days. Intestinal stem cells are located at the bottom of the intestinal crypt. They divide, amplify, and flow onto the villi, where they differentiate, absorb nutrients, and eventually die at the villus tips. The stem cells can differentiate into absorptive enterocyte and multiple secretory cells such as paneth cells, goblet cells, and enteroendocrine cells.

1.1.2 The Location of the Skeletal Muscle Stem Cell Niche

In skeletal muscle, stem cell is surrounded by the myocyte under the basal membrane. The satellite cell is slow turnover and begin to differentiate when it was stimulated by injury and demand.

1.2 Components of the Niche

The niche is a complex and dynamic structure that transmits and receives signals through cellular and acellular mediators. Hypothetical niche covers known components of previously described mammalian and non-mammalian niches: the stem cell itself, stromal cells, soluble factors, extracellular matrix, neural inputs, vascular network and cell adhesion components. It is important to note that although many niche components are conserved, it is unlikely that every niche necessarily includes all of the components listed. Instead, niches are likely to incorporate a selection of these possible avenues for communication, specifically adapted to the particular functions of that niche, which might provide structural support, trophic support, topographical information and/or physiological cues[1].

1.2.1 Niche Cells

The stem cell niche contains a variety of cell types, each with a distinct function. At present, it is clearly illustrated in the case of the hematopoietic microenvironment localized in the space of adult bone marrow. It comprises a range of different cell types including osteoblastic, vascular and neural cells, megakaryocytes, macrophage and immune cells, each have important roles and can be considered to regulate distinct HSC niches. Wnt and Notch signals promote the proliferation of HSCs (Fig. 4-1).

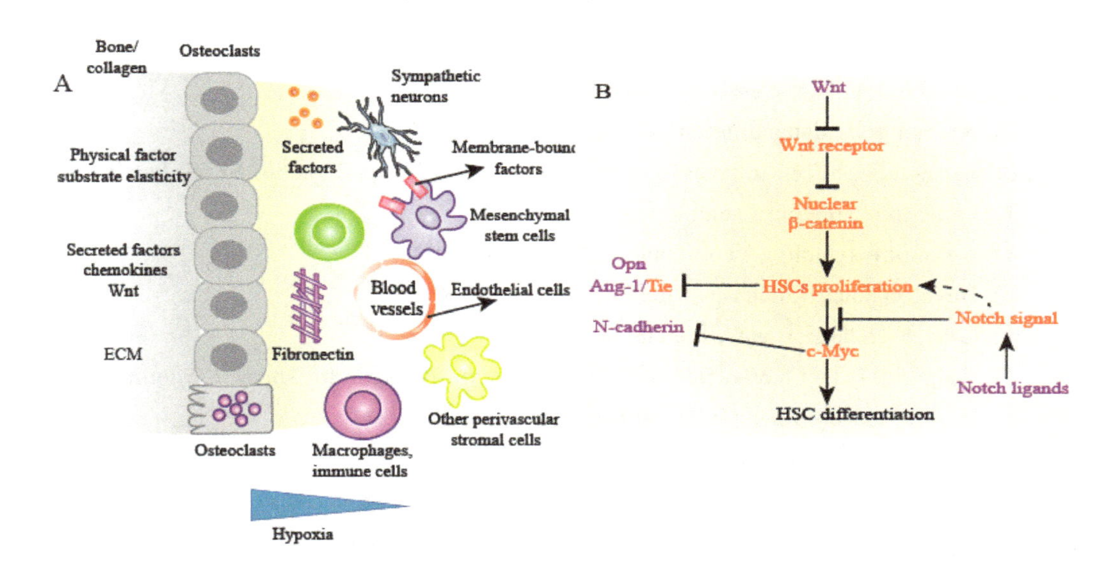

Fig. 4-1 The hematopoetic stem cell niche

Illustrated by 张伟娴.

Reference:

LANE S W, WILLIAMS D A, WATT F M. Modulating the stem cell niche for tissue regeneration [J]. Nature biotechnology. 2014, 32 (8): 795-803.

In intestinal stem cell niche, key niche cell types including the differentiated progeny of the stem cells. Paneth cells physical co-localize with and support intestinal cells through secretion of Wnt, Notch and EGF (Fig. 4 - 2).

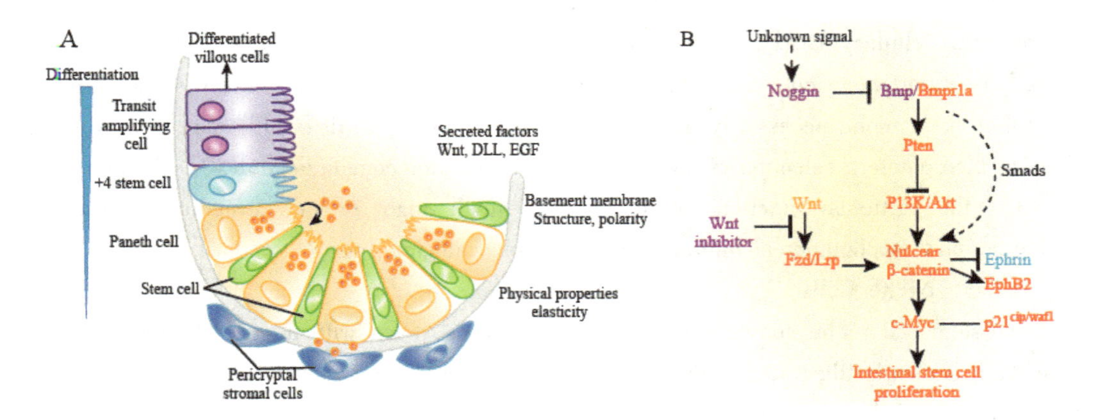

Fig. 4 - 2 The intestinal stem cell niche

Illustrated by 张伟娴.

Reference:

LANE S W, WILLIAMS D A, WATT F M. Modulating the stem cell niche for tissue regeneration [J]. Nature biotechnology. 2014, 32 (8): 795 - 803.

Stromal cells regulate stem cell function:

(1) Stromal cells depending on the location.

(2) Signals to stem cell to keep them in the niche, promote proliferation or differentiation.

(3) Paracrine signaling (soluble cues).

(4) Juxtacrine signaling (cell contact dependent cues).

(5) Direct contact through gap junctions.

(6) Cell-cell adhesion via adherens junctions.

(7) Cadherins: HSCs associate with osteoblasts via N-cadherin, muscle satellite cells associate with the muscle fiber via M-cadherin.

1.2.2 Secreted Factors

The communication between stem cell and niche stromal cell is mediated by secreted factors. Cell signals and systemic factors affect the proliferation and differentiation of stem cells in the form of autocrine or paracrine.

1.2.3 Cell Adhesion Molecules

Ensure that stem cells settle in the microenvironment and are regulated by signal molecules. The most important cell adhesion molecules are integrins.

1.2.4 Extracellular Matrix

The extracellular matrix (ECM) is the key component of stem cell niche presenting within all tissues and organs in human body. The essential components of all ECM are water, proteins (different fibers: collagens, elastin, laminin, fibronectin, etc.) and polysaccharides that are produced and organized by cells and secreted proteins (ECM metalloproteases). However, their composition, architecture and bioactivity vary considerably from tissue to tissue in relation to the specific role of the ECM which is required to assume. In general, ECM provides support including structural support for scaffolding cellular constituents and biochemical and biomechanical support for those events leading to tissue morphogenesis, differentiation and homeostasis, segregate tissues and participate in cellular communication.

In the stem cell niche, the ECM provides structure, organization and important signals for maintaining the normal function of stem cells and can directly regulate the differentiation direction of stem cells. In other words, ECM not only anchors stem cells but also directs their fate. It has been illuminated that many of the intracellular signaling pathways are involved in ECM-stem cell interactions.

1.2.5 Spatial Effect

The spatial structure plays an important role in maintaining an appropriate number of stem cells and the directional differentiation of stem cells.

1.3 Niche Factors

Stem cell niches are complex, heterotypic, dynamic structures. The stem cells are regulated by many niche factors including other cells, secreted factors, metabolic factors (oxygen, glucose, calcium), physical factors, inflammation, hypoxia, and ECM[2].

1.3.1 Secreted Factors

Indirect communication between stem cells and niche cells is mediated by secreted factors. In the hematological system, hematopoietic stem cells (HSCs) are regulated by signals from the bone marrow (BM) niche, which keep hematopoiesis at steady state and in hematologic disorders[3]. Manipulation of the HSC niche by targeting the specific signals *in vivo* is routinely exploited in clinical practice. For example, cytokine growth factors or by blocking adhesion molecules regulate HSCs mobilization. Using cytokines such as granulocyte colony-stimulating factor (G-CSF) or granulocyte-macrophage colony-stimulating factor (GM-CSF), which promotes the mobilization of HSCs from their niche, is a support treatment of hematological malignancy, BM failure and rare genetic disorders[4].

Wnt ligands have been identified as key signal proteins in both embryonic and adult stem cell niches including the intestine, liver, skin, brain, prostate, and mammary gland. The Wnt signaling pathway is involved in cell proliferation, differentiation, migration, morphological changes, and apoptosis. Wnt molecules are lipid modified making them highly insoluble. It contributes to pluripotency and stem cell self-renewal through activation of downstream signaling cascades including the Wnt/β-catenin pathway. Furthermore, Wnt proteins are often presented to recipient cells in a spatially restricted manner, which means that Wnt proteins typically act locally within a one- or two-cell diameter. Wnt signal is activated when Wnt ligand bind to the Frizzled/LRP5/6 receptor complexes at the plasma membrane, causing the stabilization and subsequent translocation of β-catenin to the nucleus where it binds to transcription factor TCF/LEF to stimulate transcription of Wnt target genes.

In niche, modulating the Wnt activity has intrinsic effects both on those cells and neighboring cells. For example, activation of the Wnt pathway in epidermal stem cells can promotes hair follicle differentiation[5], stimulates melanocyte differentiation[6] and reconstructs adult dermis to acquire characteristics of neonatal dermis[7]. Different levels of Wnt pathway activation have different effects, both within the epidermis and in the underlying dermis.

1.3.2 Wnt signaling and intestinal stem cell

In stem cell niche, four well-characterized signaling pathways tightly control the intestinal homeostasis. They include Wnt, epidermal growth factor (EGF), Notch and bone morphogenetic protein (BMP) signaling pathways. Three signals (Wnt, Notch, and EGF) are essential for intestinal epithelial stemness, whereas BMP negatively regulates stemness[8].

Wnt is the key signal to maintain stem cell fate and promote proliferation. In recent years, Wnt/β-catenin signaling has been shown to be essential for the self-renewal of a variety of mammalian stem cells. In the intestinal stem cell niche, Wnt is secreted from Paneth and stromal cells and direct to the stem cells. Intestinal stem cells receive Wnt signals from Paneth cells by direct contact. Wnt signal is activated when Wnt ligand bind to the Frizzled/LRP5/6 receptor complexes at the plasma membrane, causing the stabilization and subsequent translocation of β-catenin to the nucleus where it binds to transcription factor Tcf4, thus activating a genetic program that supports stemness. Notch signal is also important to maintain the undifferentiated state of stem cell. The Dll1/4 on Paneth cell bind to Notch on stem cell, leading to the release of Notch intracellular domain which subsequently interacts with the nuclear effector RBM-J to keep stemness state. EGF signals are essential for mitogenic effects of stem cell and transit-amplifying cell (TA cell) through binding to EGF receptors (EGFR). Consistently, the Ras/Raf/MEK/ERK signaling pathway is active in crypt. In mouse models, BMP-4 exclusively express in the intra-villus mesenchyme. Villus epithelial cells respond to the BMP signal. When BMP signals in the villus is inhibited by transgenic expression of noggin, it will format numerous ectopic crypt units perpendicular to the crypt-villus axis. BMP signals bind to BMP receptors lead to complexes between Smad1/5/8 and Smad4 to repress stemness genes in the nucleus.

1.3.3 Extracellular Matrix

In general, ECM provides structure, organization and important signals for maintaining the normal function of stem cells and can directly regulate the differentiation direction of stem cells.

(1) ECM and signaling.

A. ECM direct signaling through mechanoreceptors.

a. Integrins link the cytoskeleton to the ECM.

b. Mechanosensitive ion channels regulate ion concentration inside cells.

c. Mechanosensitive phospholipids and lipases can activate downstream signaling.

B. Binds to factors such as morphogens (e. g. BMP, Wnt) and growth factors (e. g. TGF-β, FGF), which limits the diffusion, controls spatial concentration and activity.

(2) Matrix elasticity.

Matrix elasticity affects stem cell differentiation. For example, when stem cell seed on stiff matrix, the force required to break integrin adhesion to matrix is high. Non-muscle myosin Ⅱ expression increases, myosin generated tension is withstood, focal adhesion maintained and stress fiber form. Cells osteogenically differentiate. On the contrary, when stem cell seed on soft matrix, the force required to break integrin adhesion to matrix is low. Non-muscle myosin Ⅱ expression does not increase and focal adhesion do not withstand tension. Integrins are internalized by caveolae and BMP receptors internalize with the integrins and BMP signaling decreases. Finally, cells neurogenically differentiate.

(3) Integrins in ECM mechanoregulation.

A. Integrin adhesome: the assembly of proteins recruited to the integrin—mediated cell—matrix adhesions—focal adhesions.

B. Mechanosensing: force-induced conformational changes in mechanosensitive proteins that are subjects to molecular force.

C. Mechanotransduction: mechanical stimuli are converted into biochemical signal, enabling cells to adapt to their physical surroundings ("outside-in"); cells generate traction forces using their actin myosin Ⅱ network, transmit them to focal adhesions and pull on the ECM ("inside-out").

1.3.4 Hypoxia

Low oxygen (O_2) tensions (hypoxia) occur naturally in developing embryos, maintain undifferentiated states of embryonic, hematopoietic, mesenchymal and neuronal stem cells phenotypes, and also influence proliferation and cell-fate commitment. Cells respond to their hypoxic microenvironment by stimulating hypoxia-inducible factors, which then coordinate the development of the blood, vasculature, placenta, nervous system, etc. Hypoxia can also mobilize stem cells such as MSCs, HSCs, CSCs, NSCs.

1.3.5 The Dynamic Niche: Inflammation

In normal condition, there is few immune cells such as macrophage, T cell in stem cell niche. However, in the case of tissue damage, innate and adaptive immune cells will migrate into niche in a transient fashion, to protect against pathogens or to promote wound healing. Compared to the permanent residents including endothelial cells, nerve cells and connective tissue fibroblasts, immune cells are just like the "visitors".

In niche, immune cell can regulate or promote stem cell function. In the case of hematologic malignancies, BM or HSC transplantation is used as an effective and curative therapy. One major problem for the therapy is immune resistance after transplantation. Genetically modified HSCs can drive tolerogenic expression of antigens, thereby improving the long-term efficacy of HSC transplants[9]. For acquired aplastic anemia, an organ-specific auto-immune disease characterized by pancytopenia and hypoplastic BM, immunosuppression with anti-thymocyte globulin (ATG) and cyclosporine A is an effective and safe therapy for patients without undergoing HSC transplantation[10].

Stem cells reside in a specialized regulatory microenvironment or niche. On one hand, they receive appropriate support for maintaining self-renewal and multi-lineage differentiation capacity. On the other hand, the niche may also protect stem cells from environmental insults including cytotoxic chemotherapy and perhaps pathogenic immunity. The testis, hair follicle, BM and placenta are sites of residence for stem cells and are immune-suppressive environments, called immune-privileged sites, where multiple mechanisms cooperate to prevent immune attack, even enabling prolonged survival of foreign allografts without immunosuppression. Research shows that *in vivo* imaging of regulatory T (Treg) lymphocytes providing immune privilege to the HSC niche. And this finding is being exploited in clinical trials to prevent rejection of transplanted organs[11]. In acute brain injury, the death of neuronal cell, niche-resident endothelial cells and macrophages can lead to generation of reactive oxygen species (ROS), so administration of the ROS scavenger, glutathione, promotes meningeal macrophage survival, reduces inflammation and ameliorates brain injury[12].

1.3.6 The Basal Membrane

Basal lamina lies beneath all epithelia cell layers and surrounds extracellular muscle cells and endothelial cells. The basal membrane is composed of collagen, laminin, fibronectin, nidogen, and proteoglycan. In the niche, stem cells are in contact with the basement membrane that can stop cell migration and provide polarity/orientation.

1.3.7 Influence of Mechanics on Organ Development

(1) In hematopoetic system, blood flow switches on the formation of blood cells.

(2) Fluid shear and mechanical strain affect angiogenesis formation of new vessels, vessel integrity and smooth muscle cell differentiation.

(3) Lung physical stress induces mutation of the lung tissue and differentiation of lung smooth muscle cells. It is important for bronchial development, first breath at birth induces physical changes induces mutation of alveolar structures.

(4) Gastrointestinal system: mechanical pressure influences proliferation of intestinal progenitors.

(5) Mechanical loading influences formation of cartilage, bones and joints.

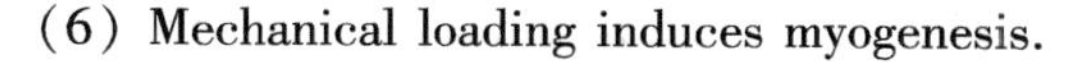

(6) Mechanical loading induces myogenesis.

1.4 Recap

(1) Stem cell niches are complex, interactive structures that integrate local and systemic signals for the regulation of stem cell activities in a spatially and temporally defined manner.

(2) Anatomically defined region.

(3) The component "part list" for niches is extensive, including cellular and acellular entities, soluble and membrane-bound signaling molecules, mechanical and chemical inputs.

(4) Complex regulation between different cell types.

(5) Signaling mechanisms are often shared between niches.

(6) Regulate stemness, prevent over proliferations, and allow differentiation when needed.

2 Introduction to Tissue Engineering for Regeneration Medicine

Regenerative medicine has been defined as the process of creating living, functional tissues to repair or replace tissue or organ function lost due to age, disease, damage or congenital defects. The field of regenerative medicine covers stem cell technology, tissue engineering and genetic engineering, and many other modern bioengineering technologies, trying to explore the possibility of tissue and organ regeneration, repair and functional reconstruction at all levels. Thus, it holds considerable promise for treating diseases that are currently intractable. Due to the extensive ability to self-renew and to generate the differentiated cell types that are needed to carry out specialized functions, stem cells are the focus of many applications in regenerative medicine. Many researchers adopt stem cell transplantation for tissue repair, alternative approach to therapy is to manipulate the stem cell microenvironment or niche to facilitate repair or regeneration by endogenous stem cells (i. e. engineering stem cell niche). For example, BM or HSC transplantation is used as curative therapy for hematologic malignancies.

Although the stem cells are just a tiny group of cells in human body, they have a tremendous impact on the biology of multicellular organisms. An improved understanding of stem cells and regenerative biology, as well as a better control of stem cell fate, is likely to produce treatments for many devastating diseases and injuries.

2.1 Manipulating of the Niche for Regenerative

The niche is highly dynamic, with multiple opportunities for intervention, so the scientists consider targeting the stem cell niche to modulate individual or multiple components of the niche to facilitate regeneration and tissue repair of damaged or diseased by activating or even manipulating normal stem cell function. These include administration of small molecules, biologics or biomaterials that target specific aspects of the niche, such as cell-cell and cell-ECM interactions, to stimulate expansion or differentiation of stem cells, or to cause reversion of differentiated cells to stem cells. This is what we call engineering stem cell niche.

At present, the main challenge in targeting the niche therapeutically is how to achieve specificity of delivery and responses. It is difficult to engineer the right niche for tissue formation. The successful treatments in regenerative medicine will involve different combinations of factors to target stem cells and niche cells, applied at different times to effect recovery according to the dynamics of stem cell-niche interactions.

The process of bioengineering: engineering the right niche for tissue formation.

(1) Identify stem cell factors and gain a deeper understanding of the molecular mechanism which dictates stem cell function.

(2) Develop the effective approach (protocols) for isolation and enrichment of stem cell population. The sources of stem cell: ESCs, adult stem cells and iPSCs.

(3) Develop biomaterials.

(4) In situ engineer to kick off the internal regenerative capacity.

2.2 The Design of Stem Cell Niche

The niche is the microenvironment that regulates and determines stem cell survival, self-renewal and differentiation. Coordinated with cell-ECM interaction, cell-cell contact, soluble factors, and physical factors (such as pH, mechanical, and electrical stimuli), defines a local biochemical and mechanical niche with complex and dynamic regulation of stem cell behavior. In order to establish a functional stem cell niche, we need to consider four aspects: signals that emanate from niche, the spatial relationship between stem cells and support cells, adhesion between stem cells and support stromal cells and/or the ECM anchors stem cells in close proximity to factors, and the structural orientation to polarize stem cells.

Decellularized tissue matrices and synthetic polymer niches are being used in the clinical trial, and people begin to know about how stem cells contribute to homeostasis and repair, for example, at sites of fibrosis. The goal of engineering stem cell niche is to develop novel approaches to build predictive models, and, ultimately, to enhance stem cell integration *in vivo* for therapeutic benefit.

2.3 Synthetic Niches *in Vivo*

Originally, many researchers adopt stem cell transplantation for tissue repair or reconstruction. However, it is clear that many niche factors such as paracrine effect, ECM stiffness and secreted factors are also important for stem cell function. An appealing approach to deal with some of the challenges in stem cell transplantation is to develop and use biomaterials that can create specialized niches for cells. Simple infusion or injection of stem cell may cause poor delivery and retention of cells at the target site or even cell death due to the loss of anchorage.

Material from nature or chemical synthesis are currently developing as vehicles to fill a specific anatomic defect to localize transplanted cells and serve as a scaffolding for formation of new tissue. It provides adhesion for recruit or disperse stem cells that improves transplanted cell survival and participates in tissue regeneration (Fig. 4 – 3). Thus, the material properties: elasticity, cell binding molecules growth factor and morphogens, need to meet criteria.

Indirect communication between stem cells and niche cells is mediated by secreted factors. Using a controlled release particle system or binding to scaffolds, the growth factors or cytokines can be specifically released into synthetic niches. For example, coordinated release of insulin-like growth factor from the transplantation vehicle can dramatically improve the cardiomyocyte function. With the release of TGF-β family proteins, bone formation by mesenchymal stems are greatly promoted.

However, full regeneration is mechanical, vascular, and neural integration of the regenerating and surrounding tissues. Materials carry angiogenic factors or ESC-derived endothelial cell progenitors can enhance local vascularization and form new vascular networks. Moreover, nervous system integration can also be enhanced by materials that provide gradients of neurotrophic factors[13].

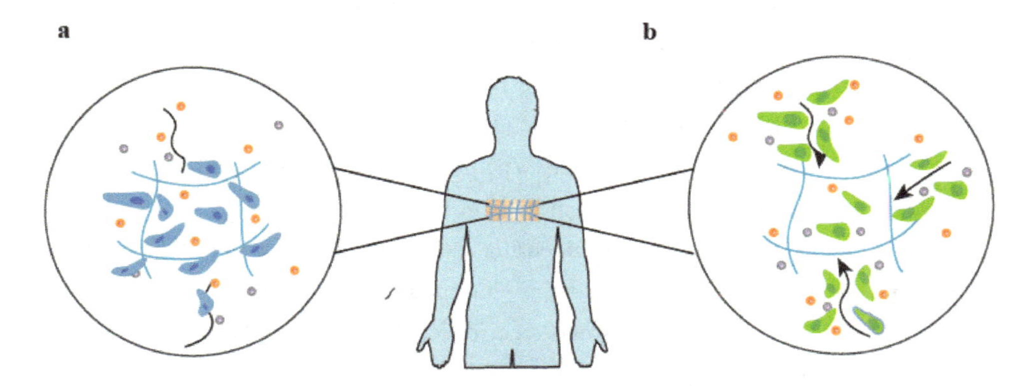

a. Dispersion of stem cells from a niche into regenerating. b. Recruitment of host stem cells for regenerating.

Fig. 4 – 3 Synthetic niches *in vivo*

Illustrated by 张伟娴.

Reference:

DISCHER D E, MOONEY D J, ZANDSTRA P W. Growth factors, matrices, and forces combine and control stem cells [J]. Science, 2009, 324: 1673 – 1677.

2.4 Ligand Immobilization

2.4.1 Immobilization of Wnt Protein Act as a Stem Cell Niche for Tissue Engineering

It is clear that Wnt signals is vital to regulate the stem cell. In vitro, there are powerful genetic tools that can be used to manipulate Wnt signaling. Wnt molecules are lipid modified making them highly insoluble. However, studies of hydrophobic Wnt proteins have been hampered by the technical challenges of purification and by their localized action. Traditionally, research has focused on methods of activating the Wnt pathway in a non-targeted manner. This includes adding small regulatory molecules, which activate Wnt signaling downstream of receptor binding, such as CHIR990021 and BIO, GSK-3b inhibitors[14] and IWR, a stabilizer of Axin and the destruction complexes[15]. Nanoparticles that bind to antibodies targeting the Frizzled2 receptor have also been used to stimulate Wnt signaling through a subset of receptor complexes. All the methods have proved to be useful, but there are still some problems such as only activating parts of the pathway or producing off-target effects including the activation of other pathways. Recently, advances in the purification and delivery of Wnt ligands have made it possible to spatially control Wnt signaling and more faithfully replicating what occurs *in vivo*. These advances provide a deeper understanding of Wnt activation at the single cell level and can be used in tissue engineering and regenerative medicine applications.

2.4.2 Non-Directional Presentation of Wnt Ligands

At the beginning, Wnt-conditioned media produced by Wnt secreting cells was used to activate Wnt signaling in cell culture. However, there are several limitations for this application: ① Wnt ligands are presented to the responsive cells in a non-directional manner; ② Wnt-conditioned media contain other secreted molecules, which may affect cellular responses; ③ Lack of precise control over the Wnt protein concentration, which is important for the cellular response and expanding stem cells *in vitro*.

With the development of protein engineering technology, purification of Wnt proteins has a breakthrough in the field. We can use these purified ligands to activate Wnt signaling in Wnt-responsive stem cells such as intestinal stem cells, HSCs and ESCs. Purified recombinant Wnts need to store in detergent to maintain activity and solubility, but it may be toxic to some cells (Fig. 4-4a). Therefore, lipid-based systems such as liposomes have been used to solubilize Wnt for delivery *in vitro* and potentially *in vivo* without the use of detergents. As the lipid moiety shield within the bilayer of the liposomes, Wnt ligands can retain their proper protein folding and biological activity (Fig. 4-4b). In addition, nanodiscs (comprise a phospholipid bilayer and an apolipoprotein A-I scaffolding component) is also used as a carrier to delivery murine Wnt3a. Wnt ligands bind to the lipid surface of the nanodisc, with the palmitoleate group inserted into the lipid bilayer (Fig. 4-4c).

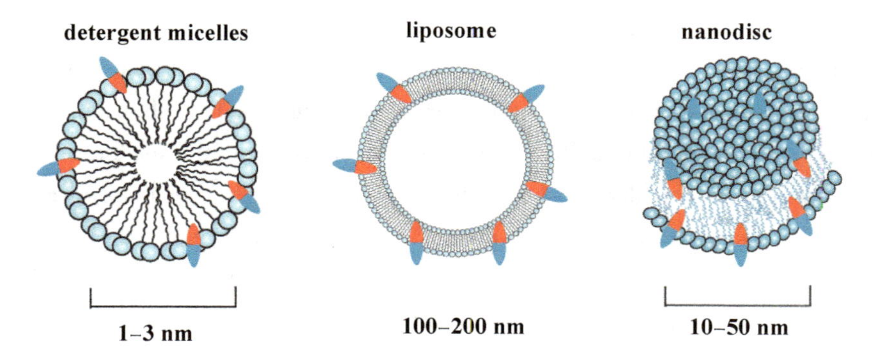

Wnts can be solubilized by detergent micelles, or carried on liposome, or nanodisc. Hydrophobic moisty of Wnts is represented in the schematic by red.

Fig. 4 -4 The non-directional delivery of Wnts

Illustrated by 张伟娴.

Reference:

MILLS K M, SZCZERKOWSKI J L A, HABIB S J. Wnt ligand presentation and reception: from the stem cell niche to tissue engineering [J]. Open biology. 2017, 7 (8): 170140.

Through the lipid-based delivery systems, Wnt can solubilize and simultaneously maintain its biological activity. However, these methods also have a major drawback. Introducing lipid biomolecules may affect cellular responses. For example, although an empty nanodisc vector has no effect on stimulating Wnt signaling, it can induce HSPC cell proliferation and expansion.

2. 4. 3 Constructing a Localized Wnt Signals Niche on Synthetic Surfaces

In vitro, presenting Wnts signals in a localized manner can better mimic cellular niches and provide possibilities to investigate how cells respond to the signals. Hence, researchers want to develop methods to achieve this through the immobilization of an active Wnts signal to synthetic surfaces. These immobilization techniques can provide a sustained, covalently bound and active Wnts signal.

There are several successful methods of immobilization of Wnts on synthetic surfaces to mimic localized Wnt presentation within the stem cell niche. Wnt covalent bound to synthetic surfaces does not disrupt the tertiary structures, in particular the disulfide bridges, is essential for maintaining signaling activity. Microbeads coated with carboxylic acid can be converted to a succinimide ester in acidic pH to facilitate covalent binding of Wnt to the bead. Besides, glutaraldehyde-coated surfaces can also be used to immobilize Wnt through a reaction of the nucleophilic groups on Wnt proteins such as amine, thiol, phenol and imidazole. Under neutral pH conditions, oligomeric hemiacetal or monomeric hemiacetal forms of the glutaradehyde can react with the functional group on the surface and/or protein. Other reactions include Schiff-base reaction and Michael-type reaction.[16]

2.4.4 Examples

Researchers adapt covalently binding to immobilized Wnt3A molecules onto commercially available aldehyde-functionalized surfaces in a one-step reaction. Recombinant Wnt3A protein bound effectively to the aldehyde surface with on average 76% of the protein remaining on the surface. By calculating, 4×10^9 Wnt3A molecules/mm^2 are immobilized onto the aldehyde surface when 20 ng of Wnt3A protein are added onto a circle with a diameter of 9 mm. After the immobilization of Wnt3A molecules, the author seeded TCF-luciferase reporter cell line (LS/L) onto the surfaces to test whether the immobilized Wnt3A remained biologically active on the surface. LS/L cells showed a dose-dependent response to increasing amounts of Wnt3A on the surface, which were all significantly higher than a surface inactivated by treatment with DTT (breaks the crucial disulfide bonds in Wnt3A). Wnt3A maintain better bioactivity because of low level of detergent (0.006%) that is harmful to protein bioactivity. This method of immobilization improves the long-term sustainability of the Wnt signaling potential while providing a localized basal source to cells. The immobilized Wnt3A surfaces can be used to induce Wnt signaling in a variety of stem cell cultures[17].

2.5 Tissue Engineering

2.5.1 Bone Tissue Engineering

Bone defect has always been a difficult problem for orthopedic surgeons. Currently, autogenous bone graft or grafting materials are employed in surgical intervention after bone tumor removal or trauma causing the defect of bone structural integrity. Autogenous cancellous bone (taken from the iliac crest, rib, fibula, or tibia) is considered to be the ideal graft material due to its biocompatible, non-immunogenic and osteogenic potential due to the presence of viable osteoprogenitor cells. Moreover, it will not transmit disease to the recipient. However, because of severely damaged and deficient, bone extracted from the patient or donor is often unable to meet the requirements of bone transplantation in the replaced parts. In addition, obtaining cancellous bone will imposes potential complications and pain to the patient that may be harmful to the patient. Xenotransplantation may lead to immune rejection. Therefore, people resort to develop new material to repair bone tissue. Bone tissue repair is a great challenge of regenerative medicine.

The basic organization of bone tissue requires the design and fabrication of a porous 3D structure or "scaffold" to contain the bone-forming cells (Fig. 4 – 5). Scaffold is complex and can be employed as bone graft substitutes for bone repair. It intends to mimic the native *in vivo* microenvironment and demands bioactivity that are also capable of supporting vascularization as well as cell proliferation and osteogenic differentiation. 3D bioactive scaffolds containing committed osteoprogenitors can provide a promising surgical tool for bone tissue engineering in clinical applications[18].

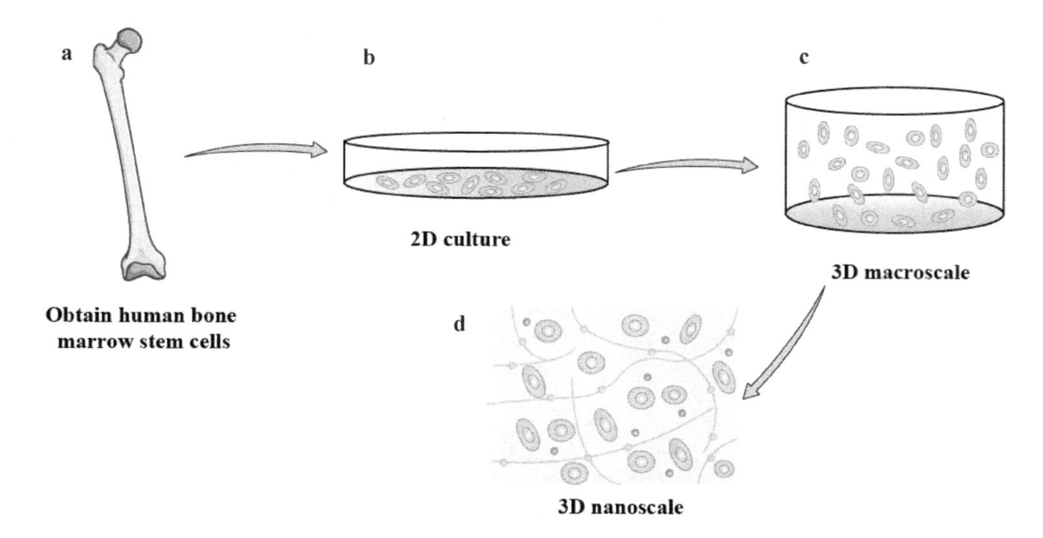

a. Obtain human bone marrow stem cells. b. 2D culture system. c. 3D culture system. Scaffold is constructed at the macro- and nanoscale levels to support individual cells and to enable whole tissue renewal. d. Nano-architecture of 3D composite scaffold. The blue circles represent adhesion motifs and the red circles represent signaling molecules.

Fig. 4 – 5 Schematic workflow of bone tissue engineering

Illustrated by 张伟娴.

Reference:

SROUJI S, KIZHNER T, LIVNE E. 3D scaffolds for bone marrow stem cell support in bone repair. [J]. Regenerative medicine, 2006, 1 (4): 519 – 528.

2.5.2 Cardiovascular Tissue Engineering

Currently, there are growing need for available substitutes of cardiovascular tissues, such as heart valves and small-diameter blood vessels. In situ, tissue engineering is emerging as a disruptive new technology (Fig. 4 – 6), which refers to biomaterial-induced endogenous regeneration directly in the injury site, or in situ, starting from readily-available, resorbable grafts that gradually transform into an autologous, homeostatic replacement tissue with the ability to repair, remodel and grow. It provides a biodegradable and cell-free constructs which are designed to induce regeneration by implantation in the functional site directly. The scaffold is 3D structure, providing the microenvironment that recruit and harbors host cells including immune cells, stem/progenitor cell and tissue cells, modulating the nature inflammatory response, and acting as a temporal roadmap for new tissue to be formed.

Tissues regeneration is based on the notion that the natural inflammatory response can be harnessed to induce endogenous tissue regeneration. The resorbable immunomodulatory scaffold provides a temporary microenvironment, which functions as an instructive road map for endogenous cells to infiltrate and create new, living, and functional tissue[19].

Although the technology has a wide application prospect, there are still many problems to be solved.

(1) How the scaffold recapitulates the complex layered architecture of cardiovascular tissues?

(2) How to predict inflammatory-driven functional regeneration?

(3) How to control biomechanical stimuli on the organization of new tissue?

(4) How fast and by which mechanism should a scaffold degrade?

(5) Is this patient-dependent and should this be personalized?

Therefore, there is still a long way to go before this new technology apply in clinical.

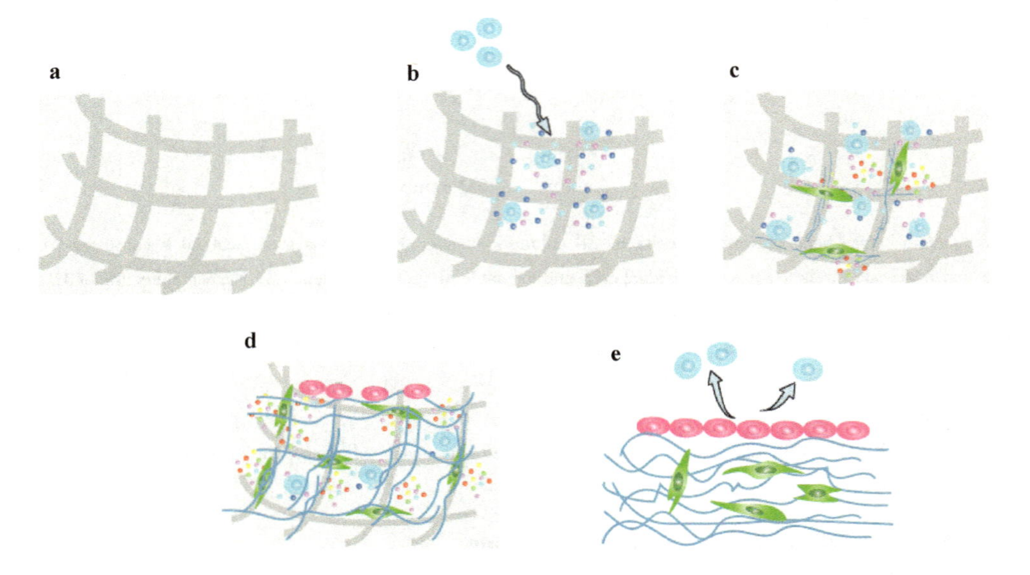

a. Scaffold synthesis. b. Host cells recruitment. c. Preliminary matrix formation. d. Scaffold biodegradation or resorption. e. Construction toward a viable substitute.

Fig. 4 – 6 Overview of the different stages of in situ tissue regeneration

Illustrated by 张伟娴.

Reference:

WISSING T T, BONITO V V, BOUTEN C C, et al. Biomaterial – driven in situ cardiovascular tissue engineering: a multi-disciplinary perspective [J]. Nature publishing group, 2017 (1): 18.

2.6 Scaffold

2.6.1 Scaffold Biomaterials Classification

2.6.1.1 Synthetic Polymers

Synthetic polymers, both organic and inorganic materials, are widely used in biomedical applications. The polymers can be biodegradable or nondegradable. For example, polylactic acid (PLA) and polyglycolic acid (PLGA) are biodegradable. In human body, PLA and PL-

GA can hydrolyze into lactic acid and glycolic acid, respectively. In addition, other biodegradable polymers and materials including polycaprolactone (PCL), polyanhydrides. Polyanhydrides are currently studied for tissue engineering applications, which has been shown to bind to soft tissue and bone.

2.6.1.2 Ceramics

Ceramics are also widely used in dental and orthopedic applications. Recently, they are being examined for bone tissue engineering applications.

2.6.1.3 Native Polymers

Native polymers such as extracellular matrix proteins are commonly exploited as scaffold in tissue regeneration. Collagens, comprising most proteins in connective tissue such as skin, bone, cartilage and tendons, are popular candidates. A variety of collagen-based products are currently under development.

2.6.1.4 Composites of Ceramics and Polymers

Composites of ceramics and polymers have the properties of both the respective materials and have been studied widely in bone tissue engineering.

2.6.2 Design Criteria

(1) Process biocompatibility that can be degraded or resorbed biologically, so that the newly formed tissues will eventually replace the scaffold.

(2) Possess appropriate interconnecting pores facilitate cells and tissue integration and vascularization.

(3) Have appropriate surface chemistry to promote cellular attachment, differentiation and proliferation.

(4) Safety. Not induce any adverse response.

(5) Easy to synthesize and fabricate into a variety of shapes and sizes.

(6) Possess adequate mechanical properties that can be implanted into repair site.

2.7 Question

Stem cell fate is regulated by a variety of niche factors including biochemical cues (growth factors, cytokines, hormones), cell-cell interaction cues (cadherins, Notch ligands), ECM components (fibronectin, laminins, collagens), physical cues (niche stiffness, pH, electricity). Do they interfere? How to test the effect of combining different niche factors on the stem cells behavior? More detail please read the articles "Stem cell niche engineering through droplet microfluidics"[20] and "High throughput screening for discovery of materials that control stem cell fate"[21].

Supplement

List of Abbreviations	
BM	bone marrow
BMP	bone morphogenetic protein
CSCs	cancer stem cells
ECM	extracellular matrix
EGF	epidermal growth factor
HSCs	hematopoietic stem cells
PLA	polylactic acid
PLGA	polyglycolic acid
ROS	reactive oxygen species
TCF	transcription factor

Key Words List	
干细胞微环境	stem cells niche
细胞外基质	extracellular matrix
再生医学	regenerative medicine
造血干细胞	hematopoietic stem cells
组织再生工程	tissue engineering

References

[1] JONES D L, WAGERS A J. No place like home: anatomy and function of the stem cell niche [J]. Nature reviews molecular cell biology, 2008, 9: 11-21.

[2] LANE S W, WILLIAMS D A, WATT F M. Modulating the stem cell niche for tissue regeneration [J]. Nature biotechnology, 2014, 32: 795-803.

[3] CALVI L M, ADAMS G B, WEIBRECHT K W, et al. Osteoblastic cells regulate the hematopoietic stem cell niche [J]. Nature, 2003, 425: 841-846.

[4] TO L B, LEVESQUE J P, HERBERT K E. How I treat patients who mobilize hematopoietic stem cells poorly [J]. Blood, 2011, 118: 4530-4540.

[5] TANIMURA S, TADOKORO Y, INOMATA K, et al. Hair follicle stem cells provide a

(To be continued)

functional niche for melanocyte stem cells [J]. Cell stem cell, 2011, 8: 177 - 187.

[6] COLLINS C A, KRETZSCHMAR K, WATT F M. Reprogramming adult dermis to a neonatal state through epidermal activation of β-catenin [J]. Development, 2011, 138: 5189 - 5199.

[7] DRISKELL R R, LICHTENBERGER B M, HOSTE E, et al. Distinct fibroblast lineages determine dermal architecture in skin development and repair [J]. Nature, 2013, 504: 277 - 281.

[8] SATO T, CLEVERS H. Growing self-organizing mini-guts from a single intestinal stem cell: mechanism and applications [J]. Science, 2013, 340: 1190 - 1194.

[9] COLEMAN M A, BRIDGE J A, LANE S W, et al. Tolerance induction with gene-modified stem cells and immune-preserving conditioning in primed mice: restricting antigen to differentiated antigen-presenting cells permits efficacy [J]. Blood, 2013, 121: 1049 - 1058.

[10] ROSENFELD S J, KIMBALL J, VINING D, et al. Intensive immunosuppression with antithymocyte globulin and cyclosporine as treatment for severe acquired aplastic anemia [J]. Blood, 1995, 85: 3058 - 3065.

[11] FUJISAKI J, WU J, CARLSON A L, et al. *In vivo* imaging of Treg cells providing immune privilege to the hematopoietic stem-cell niche [J]. Nature, 2011, 474: 216 - 219.

[12] ROTH T L, NAYAK D, ATANASIJEVIC T, et al. Transcranial amelioration of inflammation and cell death after brain injury [J]. Nature, 2014, 505: 223 - 228.

[13] DISCHER D E, MOONEY D J, ZANDSTRA P W. Growth factors, matrices, and forces combine and control stem cells [J]. Science, 2009, 324: 1673 - 1677.

[14] MYERS C T, APPLEBY S C, KRIEG P A. Use of small molecule inhibitors of the Wnt and Notch signaling pathways during Xenopus development [J]. Methods, 2014, 66: 380 - 389.

[15] CHEN B, DODGE M E, TANG W, et al. Small molecule-mediated disruption of Wnt-dependent signaling in tissue regeneration and cancer [J]. Nature chemical biology, 2009, 5: 100 - 107.

[16] MILLS K M, SZCZERKOWSKI J L A, HABIB S J. Wnt ligand presentation and reception: from the stem cell niche to tissue engineering [J]. Open biology, 2017, 7: 170140.

[17] LOWNDES M, ROTHERHAM M, PRICE J C, et al. Immobilized WNT proteins act as a stem cell niche for tissue engineering [J]. Stem cell reports, 2016, 7: 126 - 137.

[18] SROUJI S, KIZHNER T, LIVNE E. 3D scaffolds for bone marrow stem cell support in bone repair [J]. Regenerative medicine, 2006, 1: 519 - 528.

[19] WISSING T B, BONITO V, BOUTEN C V C, et al. Biomaterial-driven in situ cardiovascular tissue engineering—a multi-disciplinary perspective [J]. NPJ regenerative medicine, 2017, 2: 18.

[20] PATEL A K. High throughput screening for discovery of materials that control stem cell fate [J]. Current opinion in solid state and materials science, 2016, 20: 202-211.

[21] ALLAZETTA S, LUTOLF M P. Stem cell niche engineering through droplet microfluidics [J]. Current opinion in biotechnology, 2015, 35: 86 - 93.

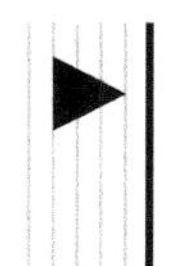

Chapter 5 Cancer Stem Cells
第五章 肿瘤干细胞

［中文导读］

传统观念认为，肿瘤细胞是由体细胞突变而来的，每个肿瘤细胞都可以无限地生长与增殖。后来，科学家们发现并非所有的肿瘤细胞都能无限增殖，只有极少数的肿瘤细胞能自我更新、增殖和诱导肿瘤形成。因此，科学家们提出了肿瘤干细胞学说。1997 年，Jhon 与 Bonnet 首次在急性髓性白血病患者中鉴定出肿瘤干细胞，这是肿瘤干细胞研究领域的一个重大突破。紧接着，科学家们又证实了在乳腺癌、肠癌、脑瘤等肿瘤组织中也存在肿瘤干细胞。有人把肿瘤干细胞比作蜂巢中的蜂后，其能够不断地产生工蜂（肿瘤细胞），扩大蜂巢，并且产生下一代蜂后。如果摧毁了蜂巢，但没有杀死蜂后，那么蜂后又会重新建立蜂巢，产生工蜂，这就好比肿瘤的复发。只要有肿瘤干细胞的存在，肿瘤就无法被彻底消除。

根据美国癌症研究协会（American Association for Cancer Research）的定义，肿瘤干细胞是肿瘤组织中一小群具有自我更新能力的细胞，可以分化成各种不同的肿瘤细胞以组成不同的肿瘤组织，从而造成肿瘤的高度异质性。肿瘤干细胞被认为是导致癌变、促进癌症进程、肿瘤转移、癌症复发和耐药性的重要原因。因此，开发有效的抗肿瘤药物，选择性地清除肿瘤干细胞或诱导肿瘤干细胞向正确的方向分化具有重要意义。

1 Introduction to Cancer Stem Cells

Cancer is the most common type of malignant tumor arising from epithelial tissue. Cancer starts when cells in a part of the body begin to grow out of control caused by genetic and epigenetic changes. It can start in any part of body such as lung, liver, breast, colon, skin, or even in the blood. Normal cells divide under precisely control of gene and are eliminated when they are aging or damaged, while cancer cells grow much faster than normal cells and proliferate without restriction, ultimately crowd out normal cells. The most common and effective treatments for cancer are surgery, chemotherapy, and radiation.

Cancer is a highly heterogeneous and developmental disease. Normal healthy cells mutate and transform into cancer cells due to accumulation of genetic mutation in oncogenes and tumor suppressor genes, as well as epigenetic alterations such as abnormal methylation and histone modification. The conventional antitumor therapy including radio- and chemotherapy that aim at eliminating tumor cells are based on these theories. However, these therapies have limited effects and poor prognosis for patients in advanced stages of cancer.

Conventional views hold that tumors are created by somatic mutations that allow each tumor cell to grow indefinitely. As the research progresses, scientists have found that not all tumor cells can proliferate without limit, and only a few tumor cells have the ability to regenerate, proliferate and induce tumor formation automatically. Hence, hypothesis of cancer stem cells (CSCs) was proposed. In 2003, Tannishtha Reya et al. published an article in *Nature* and formally put forward the contemporary cancer stem cell theory. Over the past decade of research, CSCs hypothesis has been gradually acknowledged. In 1997, J. E. Dick and colleagues identified the first CSC, the leukemic stem cells (LSCs), in samples from patients with acute myeloid leukemia (AML). [1] Both LSCs and normal stem cells express specific cell surface markers ($CD34^+/CD38^-$). Later on, several studies of mouse models of breast[2], brain[3], liver and intestinal proved that CSCs exist and arise during tumor formation. It has been confirmed that CSCs play a critical role in tumor initiation, progression, metastasis, recurrence, and resistance to radio-and chemotherapy. Therefore, it is of great significance to develop effective therapeutic strategies that can selectively eliminate CSCs or induce them into proper differentiation direction.

1.1 Definition

CSCs, as defined by the American Association of Cancer Research, are a small subset of cells with the capability of self-renewal and differentiation into the heterogeneous lineages that constitute the tumor mass.

1.2 Characteristics

1.2.1 Self-renewal

CSCs process self-renewal ability. They are resistant to apoptosis and can proliferate without restriction, which means that these cells are not subject to the aging effect and have an infinite replication potential.

1.2.2 Multipotency

CSCs can differentiate into a variety of cells, generate numerous daughter cells.

1.2.3 Resistance to Radio- and Chemotherapy

CSCs, a small subset of the cancer cell population, are responsible for the failure of existing cancer treatment because of the therapeutic resistance including radio-and chemotherapy. Thus, they have become a main target in cancer treatment. The main mechanisms of radio-and chemotherapy resistance is related to some intrinsic and acquired factors, including DNA damage repair ability, high level of ATP bind cassette (ABC) transporters, hypoxia adaption, immune evasion ability, and upregulated antiapoptotic proteins. Therefore, several strategies can be design to kill CSCs: ① destroy the CSCs niche, ② inhibit crucial signaling pathways, ③ target surface markers of CSCs, and ④ selectively inhibit ATP-driven efflux transporters. The combination of conventional therapy and targeted CSCs therapy presents better anti-tumor effect.

Although conventional radio- and chemotherapy can eliminate most of the dividing non-stem cells, surviving CSCs can repopulate the tumor. Therefore, targeting both CSCs and the dividing cells would be required for complete tumor eradication (Fig. 5 – 1).

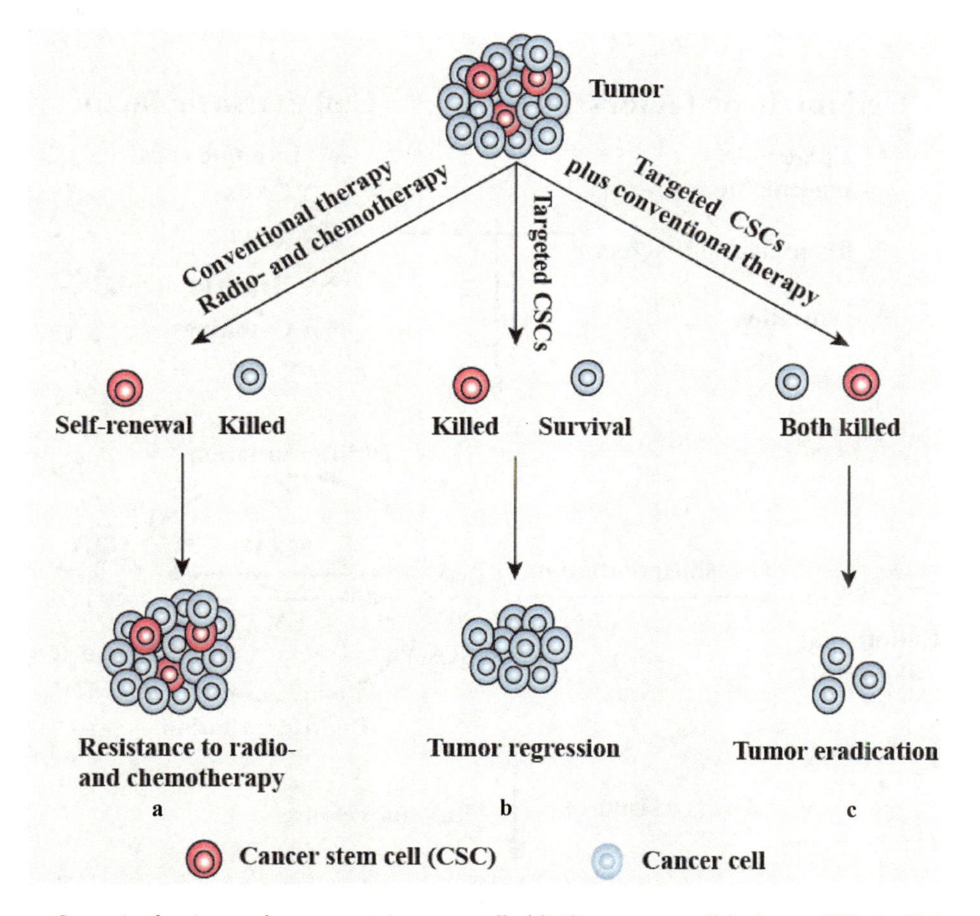

a. Conventional anticancer therapy: targeting cancer cells (dividing non-stem cells), but not CSCs. b. CSC-specific therapy: targeting CSCs. c. Combination of conventional anticancer therapy and CSC-specific therapy have the potential for tumor eradication.

Fig. 5 – 1 Anticancer strategies of targeting cancer cells or CSCs

Illustrated by 张伟娴.

Reference:

HUNTLY B J, GILLILAND D G. Leukaemia stem cells and the evolution of cancer-stem-cell research [J]. Nature reviews cancer. 2005, 5 (4): 311 – 321.

1.2.4 Metastatic

There are many reasons for recurrence and metastasis, one of which is the presence of CSCs. CSCs lead to the migration and propagation of tumor in new site.

1.2.5 Plasticity

CSCs plasticity is the ability to dynamically transit between CSCs and non-CSCs states. It is a complex process regulated by both cell intrinsic and extrinsic factors (Fig. 5-2). Plasticity plays a crucial role in the drug resistance, tumor relapse and metastasis.[4]

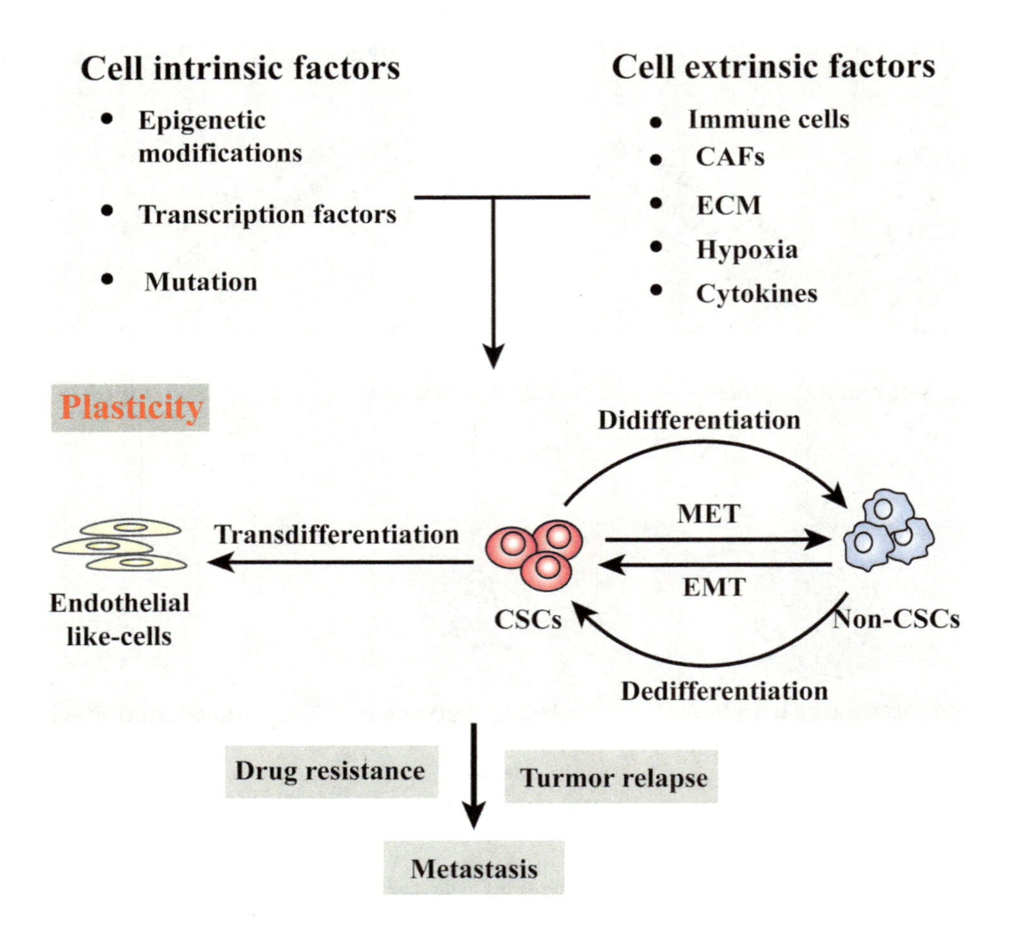

Fig. 5-2 Cancer stem cells plasticity

Illustrated by 张伟娴.

Reference:

THANKAMONY A P, SAXENA K, MURALI R, et al. Cancer stem cell plasticity-A deadly deal [J]. Frontiers in molecular biosciences, 2020, 7: 79.

1.3 Mechanism of Cancer Stem Cell Resistance to Therapy

1.3.1 Drug Efflux

Drug efflux mediated by ATP bind cassette transporters family is an important mechanism of tumor multidrug resistance. ABC transporters family comprise 49 transmembrane protein, several of them are correlated with multidrug resistance, named multidrug resistance protein 1 (MDR1; ABCB1), multidrug resistance-associated protein 1 (MRP1; ABCC1) and breast cancer resistance protein (BCRP; ABCG2).

The selective expression of several members of the ABC transporter family has been reported in some types of CSCs, such as lung cancer cell line[5]. High level of ATP bind cassette transporters (ABC transporters) efflux the chemotherapeutic drugs, causing drug resistance in CSCs.

1.3.2 Enhanced DNA Damage Response and Repair Ability

Radiotherapy and chemotherapeutic drugs are genotoxic and can damage genes. CSCs have been shown to have enhanced DNA damage response activation as a mechanism to repair DNA damage effectively.

1.3.3 Hypoxia Adaption

Hypoxia is a common characteristic of all solid tumors and is correlated with disease progression, poor prognosis, and treatment. HIF-1α and/or HIF-2α, regulators of O_2 homeostasis, play a crucial role in cancer cells survival, proliferation, differentiation, angiogenesis, metabolism, invasion, and metastasis. For example, knocking down HIF-1α and/or HIF-2α can inhibit the proliferation, induce apoptosis of human glioblastoma CSCs, and suppress tumor growth *in vivo*[6]. In melanoma cell lines, hypoxia condition induce the expression of Octamer (Oct)-4, which is involved in decreased differentiation, spheroid formation, and resistance therapy[7]. Moreover, hypoxia induce the expression of vascular endothelial growth factor (VEGF) that can promote angiogenesis, thereby helping cancer cells to obtain more oxygen and nutrients.

1.3.4 Upregulation of Antiapoptotic Proteins

Radiation and chemotherapeutic drugs can induce cell apoptotic. Apoptosis process is regulated by the balance of pro-apoptotic (such as BAD, BAX, and BAK) and antiapoptotic (such as BCL2, MCL1, and BCL-XL) members of the BCL2 family of proteins, which regulate the permeabilization of the outer mitochondrial membrane. Overexpression of antiapoptotic proteins protect cancer cells from apoptosis induced by radio- and chemotherapy.

1.3.5 Quiescence

Conventional anticancer therapy is more efficient against cells with high proliferative rates. Experimental evidence has shown that tumor containing quiescent cells, such as glioblastoma, melanoma, pancreatic cancer, lung cancer, and breast cancer, are correlated with drug resistance. CSCs, isolated from ovarian and breast tumors, are relatively quiescent[8-9]. Induction of quiescent CSCs to enter the cell cycle, followed by chemotherapy, has potential to improve drug resistance.

1.3.6 Pro-Survival Signaling

Similar to normal stem cell, cancer stem cell behaviors (e.g. survival, proliferation, and the balance between self-renewal and differentiation) are controlled by specific signaling pathways, triggered by the integration of intrinsic factors with extrinsic factors supplied by the surrounding microenvironment. There are many signaling pathways involved in the regulation of CSC (e.g. Wnt, Notch, BMP, Hedgehog, Janus kinase/signal transducers, and activators of transcription). For example, Wnt ligands have been identified as key signal proteins in both embryonic and adult stem cell niches including the intestine, liver, skin, brain, prostate, and mammary gland. The Wnt signaling pathway is involved in cell proliferation, differentiation, migration, morphological changes, and apoptosis, and has been associated with CSC activity. Kendziorra and colleagues have shown that the Wnt transcription factor 4 (TCF4) is upregulated in 5-fluoruracil-resistant colorectal cancer cell lines. Silencing TCF4 via siRNA can sensitize them to radio- and chemotherapy. [10]

1.4 Origin of Cancer Stem Cells

Currently, the source of CSCs is not totally clear and may differ depending on the specific disease. There are three acceptable sources of CSCs including normal stem cells, progenitor cells and differentiated cells.

1.4.1 Cancer Stem Cell Arise by Mutations from Normal Stem Cells

Normally, stem cell process ability to self-renew, differentiate into a variety of cells, and generate various daughter cells. A characteristic of self-renewing cells is an increase in telomerase activity. The length of the telomeres remains constant after cell division, which means that these cells can avoid aging and apparently have unlimited proliferation potential[11-12]. For differentiation potential, stem cell can differentiate into following groups: ① Totipotent stem cells are capable of giving rise to any cell type of an organ or placenta; ② Pluripotent stem cells, such as embryonic cells, are capable of giving rise to any cell type of an organ, but not placenta; ③ Multipotent stem cells, such as HSCs and MSCs, are capable of giving rise to specialized cell types present in a specific tissue or organ, most adult stem cells are multipotent stem cells; ④ Unipotent stem cells, such as ESCs and cardiac stem cell (CDCs), are capable of giving rise to only one cell type of a tissue. Stem cell self-renewal and differentiation processes are precisely controlled by niche regulatory systems.

The normal stem cells transform into cancer stem cells (CSCs) is due to the accumulation of genetic mutation in oncogenes, suppressor genes and miss-match repair genes, as well as epigenetic alterations such as abnormal methylation, histone modification. For example, experimental results confirm that acute myeloid leukemia stem cells are derived from normal stem cells that have undergone oncogenic transformation. Both LSCs and normal stem cells express specific cell surface markers ($CD34^+$/$CD38^-$)[13].

1.4.2 Cancer Stem Cell Arise from Progenitor Cells

Progenitor cells are partly differentiated, with a limited proliferation potential. CSCs arises from a committed progenitor cell that reacquired stem-cell-like properties during oncogenic transformation.

1.4.3 Cancer Stem Cell Arise from Differentiated Cells

Another source of cancer stem cells is differentiated cells. Differentiated cells dedifferentiate due to oncogenic mutations, thereby acquiring stem cell-like properties and becoming cancer stem cells.

1.5 Characterize Cancer Stem Cells

The characteristics of CSCs are shown through the following methods.

(1) Biochemical methods.

(2) Biophysical methods.

(3) Single-cell methods.

(4) Computational methods.

1.6 Cancer Stem Cells Markers

Various surface markers of cancer stem cells have been identified. Surface markers can be used to identify, enrich or isolate CSCs, so as to achieve the purpose of molecular targeted therapy. In addition, different tumors have different cell surface markers, which is beneficial for cancer diagnosis and treatment (Tab. 5 – 1).

Tab. 5 – 1 Cancer stem cell markers identified for various cancers

Tumor Type	Cell Surface Marker	References
Acute myeloid leukemia	$CD34^+$, $CD38^-$	[13 – 16]
Hepatocellular carcinoma	$CD45^-$, $CD90^+$	—
Colon cancer	$CD133^+$, $CD44^-$, $CD26^+$, ALDH	[17 – 18]
Breast cancer	EPCAM $(ESA)^+$, $CD44^+$, $CD24^-$, ALDH, CD29, CD133	[2]
Ovarian cancer	$CD133^+$, $CD44^+$, $CD117^+$, $CD24^+$	[19]
Glioblastoma	$CD133^+$, $CD15^+$	[3]
Medulloblastoma	$CD133^+$, $CD15^+$	[20]
Small cell and none-small cell lung cancer	$CD133^+$	[21 – 22]

(To be continued)

Tumor Type	Cell Surface Marker	References
Prostate cancer	$CD44^+$, $CD133^+$, CD49	[23]
Melanoma	$CD20^+$, $CD271^+$	[24]
Pancreas adenocarcinoma	$CD44^+$, $CD24^+$	[25]
Rectal Cancer	$CD133^+$	[26]
Head and neck squamous cell carcinoma	$CD44^+$, ALDH1	[27]

1.7 Isolation of Cancer Stem Cells

There are three main techniques used for the sorting of CSCs, including long term cell culture, fluorescence activated cell sorting (FACS), and magnetic bead cell sorting (MACS).

1.7.1 Long Term Cell Culture

CSCs process self-renewal ability. They are resistant to apoptosis and can proliferate without restriction, which means that these cells are not subject to the aging effect and have an infinite replication potential. By long-term culturing, CSCs can be isolated from other tumor cells which do not have self-renewal ability.

1.7.2 Fluorescence-Activated Cell Sorting Technique

This technique can isolate cells based on the expression of special proteins, epigenetic alternation and the expression of cellular surface markers like CD24, CD133, CD34, ALDH1, and CD44. For example, CSCs are labeled by using special monoclonal antibody with fluorescence. According to fluorescence intensity, CSCs can be separated from other cancer cells.

1.7.3 Magnetic Cell Sorting Technique

This technique can isolate cells based on expression of special stem cell surface markers like CD133. Before isolation, cells with CSCs markers like CD133 will be labeled using special monoclonal antibody or magnetic microbead like anti-CD133. Then the labeled cells will be adsorbed on the sorting device, and the remaining cells will be eluted. Third step is to wash down and collect the labeled cells. After positive selection, CSCs can be separated from other cancer cells. MACS is considered as the most effective and direct ways to isolate CSCs, because different types of CSCs have their special markers (Tab. 5 – 1). We can use these markers to detect and isolate CSCs.

1.8 The Correlation Between Cancer Stem Cells, Epithelial-Mesenchymal Transition and Vasculogenic Mimicry Formation

1.8.1 Vasculogenic Mimicry

As solid tumors increase in size, they can't get enough nutrients and oxygen from the local microenvironment. In response to this stress, tumor cells induce the endothelial cells of normal blood vessels to form new vessel, which is called angiogenesis (Fig. 5 - 3) . For many years, angiogenesis was considered to be the only way of tumor vascularization. Several anti-angiogenesis agents, primarily targeting endothelial cells, have been applied clinically to suppress tumor growth. However, they only temporarily slow tumor growth and the tumor often become resistant. This indicates that there may be an alternative blood supply patterns providing nutrients and oxygen to tumor cells.

a. Angiogenesis

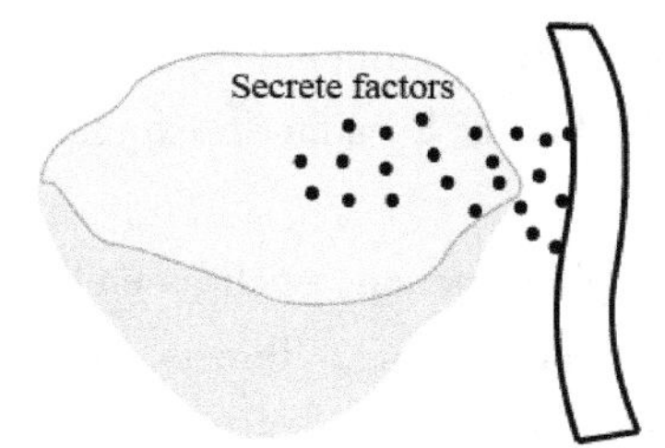

Tumor cells secrete factors to stimulate vessel growth.

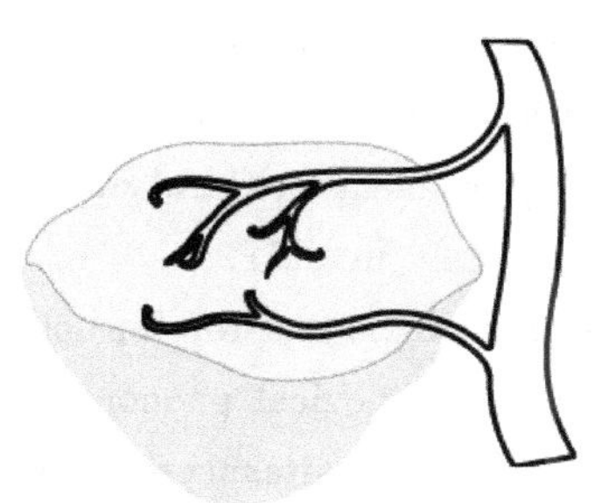

Vessel grows into the tumor mass.

b. Vasclogenic mimicry(VM)

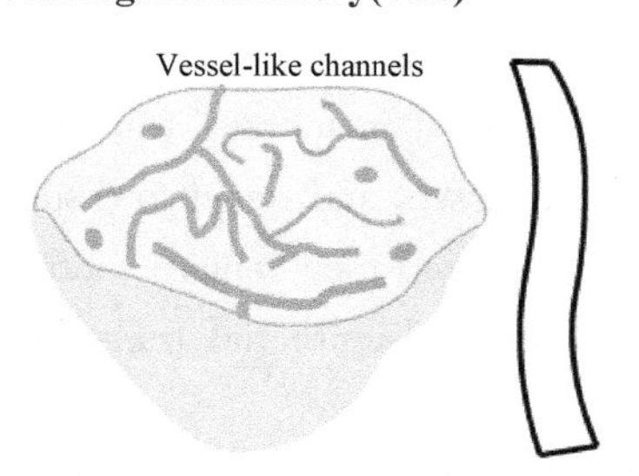

Tumor cells organize themselves to form vessel-like channels.

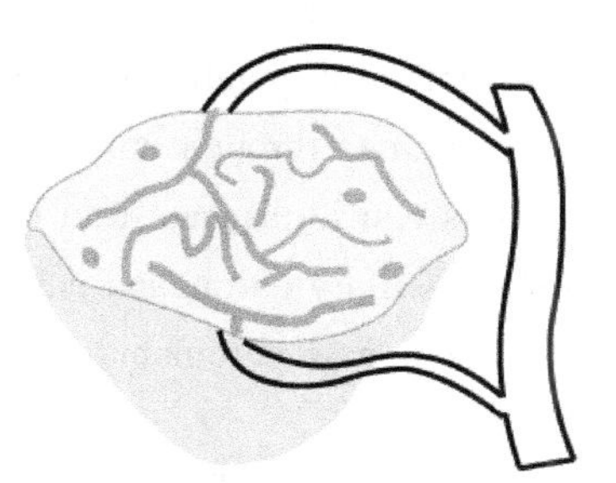

Normal blood vessels hook up to channels within the tumor mass.

Fig. 5 - 3 Two ways to feed a tumor

Illustrated by 张伟娴.

Reference:

LESLIE M. Tumors' do-it-yourself blood vessels [J]. Science. 2016, 352 (6292): 1381 - 1383.

In 1999, Maniotis et al. found that highly aggressive and metastatic melanoma cells are able to form vessel-like structure highly patterned vascular channels lined externally by tumor cells[27]. This process was termed vasculogenic mimicry (VM). Later on, VM has been observed in numerous types of aggressive tumors, such as glioblastoma, breast cancer, colorectal cancer, ovarian carcinoma, and astrocytoma. Different from angiogenesis, VM is composed of tumor cells and a basement membrane, without the endothelial cells that line normal blood vessels. It suggests that the cancer cells themselves form vessel-like channels. VM was categorized into two distinctive types: the matrix type and the tubular type. Blood plasma and red blood cells are able to flow into the channels within the tumor mass. Patients with tumors undergoing VM have a worse prognosis. Angiogenesis inhibitors can not suppress the formation of VM, and even promote the extracellular matrix-rich tubular network formation *in vitro*.

1.8.2 Epithelial-Mesenchymal Transition

Epithelial-mesenchymal transition (EMT) refers to the transfer of epithelial cells to fibroblasts or mesenchymal cells. This process is characterized by the loss of epithelial traits and the acquisition of mesenchymal phenotypes. Through EMT, epithelial cells loose cell-cell contact and acquire metastatic and invasive ability. The loss of E-cadherin expression seems to be closely involved in EMT.

ETM plays an important role in both physiological phenomenon, such as embryonic development, and physiological phenomenon, such as wound healing, fibrosis, and invasion and metastasis of epithelial cancers. Activation of EMT induces tumor cell to metastasize and invade distant organ.

Recent studies have reported that EMT involve in the formation of VM. The upregulation of EMT-associated transcription factors are observed in VM-forming tumor cells.

Normal epithelia lined by a basement membrane can transform into a carcinoma in situ due to the accumulation of genetic and epigenetic mutation. As the tumor grows, carcinoma cells begin to disseminate through an EMT, and the basement membrane becomes fragmented. Carcinoma cells can intravasate into lymph or blood vessels and travel to distant organs. At the distant site, solitary carcinoma cells can extravasate and then remain solitary (micrometastasis) or form a new carcinoma through a mesenchymal-epithelial transition (MET) (macrometastasis).

1.8.3 Cancer Stem Cells Involve in VM Formation by the Induction of EMT

Tumor tissue is highly heterogeneous, consisting different types of cells at different developmental stage. CSCs are at the top of the hierarchical pyramid. Mounting studies show that CSCs have the capacity for differentiating into tumor and endothelial lineages, as well as vascular smooth muscle-like cells. Carcinoma cells may be endowed with the stem cell phenotype through EMT. Subsequently, CSCs can generate cells with tumor and endothelial phenotype or transdifferentiate into endothelial cell to engage in the formation of VM.[28]

2 The Cancer Stem Cell Niche

2.1 Niche Components

CSCs niche is the microenvironment that regulates the fate of CSCs. Components of CSCs niche are similar to normal stem cell niche which includes ECM, niche cells, cell adhesion molecules, and signals factors (Fig. 5-4) . CSCs behavior, such as survival, proliferation, and metastasis, in particular the balance between self-renewal and differentiation, are ultimately controlled by the integration of intrinsic factors with extrinsic factors supplied by the surrounding microenvironment. [29]

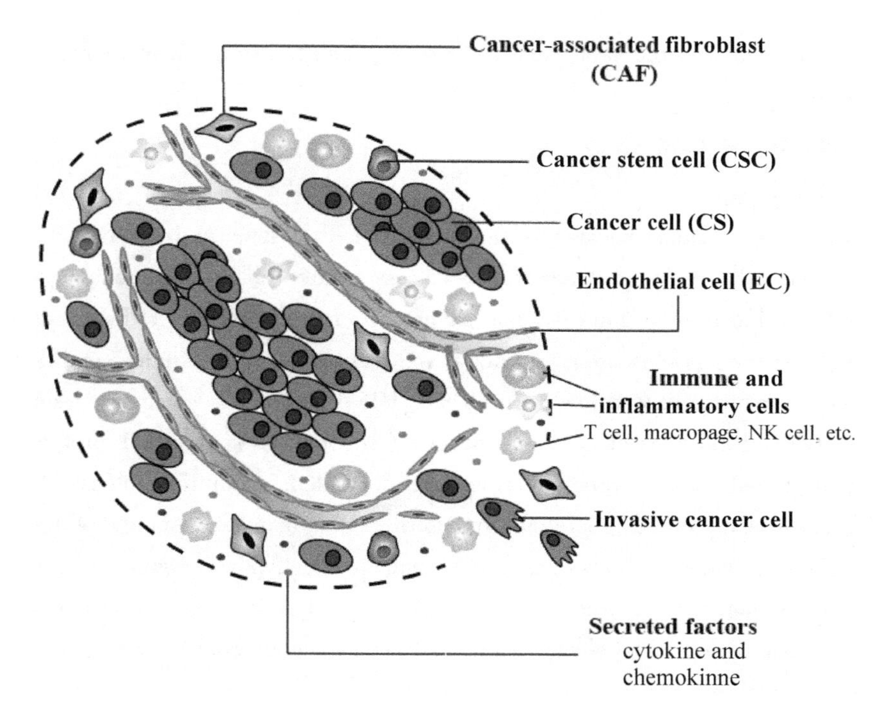

Fig. 5-4 Cancer stem cell niche

Illustrated by 张伟娴.

Reference:

HANAHAN D, WEINBERG R A. Hallmarks of cancer: the next generation [J]. Cell, 2011, 144 (5): 646-674.

2.1.1 Niche Cells

(1) Cancer cells, CSCs, invasive cancer cells.

(2) Cancer-associated fibroblast (CAF).

(3) Endothelial cells (ECs).

(4) Immune and inflammatory cells: T cell, Treg cell, NK cell, macrophage, etc.

2.1.2 Secreted Factors

The most studied signal pathway is Notch、Hedgehog and Wnt/β-catenin signaling pathways.

2.1.3 Extracellular Matrix

The ECM is an essential noncellular component of the adult stem cell niche. However, the ECM composition in the CSC niche differs from that in healthy tissue. In solid tumors, increased ECM stiffness protects CSCs from chemotherapeutic drugs.

2.1.4 Cell Adhesion Molecules

Cell adhesion molecules mediate the recognition and binding between cells, cells and ECM, and certain plasma proteins, and participate in the transmission of information inside and outside cells.

2.2 Niche Factors

2.2.1 Hypoxia

Hypoxia arise in tumors because of rapid cancer cell division and aberrant blood vessel formation. Tumor cells residing closer to blood vessels are relatively abundant in oxygen, while those far to blood vessels are hypoxic. Hypoxia-inducible factors (HIFs), consisting of HIF-α and β subunit, is ubiquitous in human and mammalian cells. Under conditions of hypoxia, oxygen levels decrease below 8% – 10%, HIF-1 proteins become increasingly stabilized, while under conditions of abundant oxygen (>8% – 10%), HIF-1 proteins express but rapidly degraded by the intracellular oxygen-dependent ubiquitin-protease degradation pathway. It can promote tumor progression by altering cellular metabolism and stimulating angiogenesis when it bound to canonical DNA sequences (hypoxia regulated elements, or HREs) in the promoters or enhancers of target genes, and activate the expression of various proteins that regulate cell metabolism, maintain basement membrane integrity, stimulate angiogenesis, and promote hematopoiesis.

In hypoxic tumor cells, stabilization of HIF-α proteins can upregulate the expression of many target genes encoding factors that mediate adaptation to hypoxic stress. Some target genes are regulated specifically by HIF-1α, such as those encoding the glycolytic enzymes ALDA and PGK, whereas others are specific targets of HIF-2α, such as those encoding tumor growth factors α (TGF-α) and cyclin D1. Most HIF target genes are regulated by both HIF-1α and HIF-2α, including those encoding the angiogenic cytokine (VEGF) and the glucose transporter 1 (GLUT1). Moreover, stabilized HIFs can enhance the expression or activity of some gene products including Notch, Oct-4, c-Myc, ABC transporters (ABC-T), and telomerase to control CSCs self-renewal and multipotency. Increased expression of Klf4, Sox2, and other factors could regulate the generation of CSCs and thus promote tumor growth (Fig. 5 – 12)[30].

In a word, hypoxia and HIFs play important roles in tumor progression.

2.2.2 Cancer-Associated Fibroblast

CAFs are important components of the tumor microenvironment, and found in nearly all solid tumors. CAFs include two distinct types: ① cells similar to the fibroblasts that create the structural foundation supporting most normal epithelial tissues; ② myofibroblasts with biological roles and properties that are different from those of tissue-derived fibroblasts. Myofibroblasts are identified because of their expression of α-smooth muscle actin (SMA). They are rare in most healthy epithelial tissues, but in several tissues, such as the liver and pancreas, there are numbers of α-SMA-expressing cells. In the case of wound, myofibroblasts transiently increase that are promote beneficial to tissue repair.

However, in the case of chronic inflammation sites, myofibroblasts can lead to the pathological fibrosis in tissues such as lung, kidney, and liver.

CAFs can enhance tumor progression, cancer cell proliferation, angiogenesis, invasion and metastasis, and development of drug resistance, through secreting of a variety of growth factors and cytokines, shaping the tumor matrix, regulating the recruitment and function of various innate immune cells and adaptive immune cells in tumor microenvironment.

2.2.3 ECM

ECM stiffness regulates tumor progression. Remodeling of the ECM has effect on tumor invasion and metastasis.

2.3 Cancer Stem Cells Derived from Mouse Induced Pluripotent Stem Cells When Exposed to a Malignant Niche

In 2012, Masaharu Seno and colleagues proved that CSCs can differentiate from normal stem cells when exposed *in vitro* to a conditioned culture medium of cancer cell lines, which is a mimic of carcinoma microenvironment. In experiment, treated with conditioned medium of mouse Lewis lung carcinoma, mouse induced pluripotent stem cells (miPSCs) acquired some characteristics of CSCs. These iPSC-derived CSCs process self-renewal ability and express the marker genes that associated with stem cell properties and an undifferentiated state. Moreover, they can form spheroids, process a high tumorigenicity in Balb/c nude mice, and exhibit angiogenesis *in vivo*[31].

2.4 Importance of Cancer Stem Cell Niche

CSCs niche hypothesis provides a novel tumor models for basic research and drug screening. Targeting the specific niche components, CSCs could be eliminated. Caution: targeting the CSCs niche can also influence the normal stem cell niche, or disrupt the levels of signals for normal cells.

(1) Target cell-cell signaling.

(2) Target stromal cells.

(3) Target ECM.

(4) Inhibit angiogenesis and formation of perivascular CSCs niche.

3 Challenges and Approaches in Cancer Stem Cells Research

3.1 Challenges

3.1.1 Small Proportion in Tumor Tissue

The proportion of CSCs in tumor tissue is very small, less than 5%. It's difficult to detect and isolate CSCs from tumor tissue or biopsy samples.

3.1.2 Lack of Specific Targets

At present, surface markers, which are identified on CSCs by experiments, are still very limited, and even express on normal stem cells as well. Hence, it is difficult to distinguish normal stem cells from tumor stem cells.

3.1.3 Complex Tumor Microenvironment

The tumor microenvironment is a complex system that is regulated by many factors such as cytokines, stromal cells, inflammation and immune response, and the extracellular matrix. Researchers want to eliminate CSCs by targeting to the specific niche components, either by depriving them of their ability to self-renew or by inducing them to differentiate normally, thereby inhibit tumor growth. However, the regulation of tumor stem cells by the microenvironment is complex and poorly understood. Secondly, CSCs niche is similar to the normal stem cells niche. Hence, there is a long way to go before developing effective therapeutic strategies that can selectively eliminate CSCs or induce them into proper differentiation direction.

3.2 Approaches in Cancer Stem Cells Research

(1) Identifying for the cell-of-origin for cancer.

A. Obtain better predictive and diagnostic.

B. Develope new and more effective targeted treatments for the disease.

(2) Discovering characteristics of normal stem cells that may become CSCs.

(3) Determining how normal stem cells form lesions that develop into cancer.

(4) Identifying markers of CSCs.

(5) Identifying molecules that can be used to stop CSCs. (Drug screening)

(6) Identifying what proteins are produced by CSCs. Are they different from those produced by normal stem cells and different types of cancers?

(7) Developing screening methods to match specific drug treatments with patients based on molecular and genetic response to the therapy.

(8) Helping the body's own cells fight cancer by engineering a cancer-fighting immune system.

Supplement

List of Abbreviations	
ABC transporters	ATP bind cassette transporters
AML	acute myeloid leukemia
BCRP	breast cancer resistance protein
BMP	bone morphogenetic protein
CAF	cancer-associated fibroblast
CSCs	cancer stem cells
CSCs	cardiac stem cells
ECM	extracellular matrix
ECs	endothelial cells
EMT	mesenchymal-epithelial transition
ESCs	epidermal stem cells
FACS	fluorescence-activated cell sorting technique
GLUT1	glucose transporter 1
HIF	hypoxia-induced factors
HSCs	hematopoietic stem cells
iPSCs	induced pluripotent stem cells
JAK/STAT	janus kinase/signal transducers and activators of transcription
LSCs	leukemic stem cells
MACS	magnetic cell sorting technique
MDR1	multidrug resistance protein 1
MET	mesenchymal-to-epithelial transition
miPSCs	mouse induced pluripotent stem cells
MRP1	multidrug resistance-associated protein 1
MSCs	mesenchymal stem cells
TCF	transcription factor
TGF-α	tumor growth factors α
TME	tumor microenvironment
VEGF	vascular endothelial growth factor
VM	vasculogenic mimicry

Key Words List	
癌相关成纤维细胞	cancer-associated fibroblast
癌症干细胞	cancer stem cells
癌症干细胞微环境	cancer stem cell niche
白血病干细胞	leukemic stem cells
表观遗传学改变	epigenetic alterations
磁珠细胞分选术	magnetic cell sorting technique
单能干细胞	unipotent stem cell
多能干细胞	pluripotent
多潜能性	multipotency
恶性肿瘤	malignant tumor
复发	recurrence
基因突变	genetic mutation
急性髓性白血病	acute myeloid leukemia
静息状态	quiescence
可塑性	plasticity
耐药	drug resistance
全能干细胞	totipotent stem cell
缺氧	hypoxia
缺氧诱导因子	hypoxia-induced factors
上皮间质转化	epithelial-to-mesenchymal transition
血管内皮生长因子	vascular endothelial growth factor
血管生成拟态	vasculogenic mimicry
抑癌基因	tumor suppressor genes
荧光激活细胞分选术	fluorescence-activated cell sorting technique
诱导多功能干细胞	induced pluripotent stem cells
致癌基因	oncogenes
肿瘤微环境	tumor microenvironment
肿瘤异质性	tumor heterogeneity
专能干细胞	multipotent stem cell
转移	metastasis
自我更新	self-renewal
祖细胞	progenitor cells

References

[1] BONNET D, DICK J E. Human acute myeloid leukemia is organized as a hierarchy that originates from a primitive hematopoietic cell [J]. Nature medicine, 1997, 3: 730 – 737.

[2] AL-HAJJ M, WICHA M S, BENITO-HERNANDEZ A, et al. Prospective identification of tumorigenic breast cancer cells [J]. Proceedings of the National Academy of Science of the United States of America, 2003, 100: 3983 – 3988.

[3] SINGH S K, HAWKINS C, CLARKE I D, et al. Identification of human brain tumour initiating cells [J]. Nature, 2004, 432: 396 – 401.

[4] THANKAMONY A P, SAXENA K, MURALI R, et al. Cancer stem cell plasticity—a deadly deal [J]. Frontiers in molecular biosciences, 2020, 7: 79.

[5] HO M M, NG A V, LAM S, et al. Side population in human lung cancer cell lines and tumors is enriched with stem-like cancer cells [J]. Cancer research, 2007, 67: 4827 – 4833.

[6] LI Z, BAO S, WU Q, et al. Hypoxia-inducible factors regulate tumorigenic capacity of glioma stem cells [J]. Cancer cell, 2009, 15: 501 – 513.

[7] KUMAR S M, LIU S, LU H, et al. Acquired cancer stem cell phenotypes through Oct4-mediated dedifferentiation [J]. Oncogene, 2012, 31: 4898 – 4911.

[8] GAO M Q, CHOI Y P, KANG S, et al. CD24 + cells from hierarchically organized ovarian cancer are enriched in cancer stem cells [J]. Oncogene, 2010, 29: 2672 – 2680.

[9] PECE S, TOSONI D, CONFALONIERI S, et al. Biological and molecular heterogeneity of breast cancers correlates with their cancer stem cell content [J]. Cell, 2010, 140: 62 – 73.

[10] KENDZIORRA E, AHLBORN K, SPITZNER M, et al. Silencing of the Wnt transcription factor TCF4 sensitizes colorectal cancer cells to (chemo-) radiotherapy [J]. Carcinogenesis, 2011, 32: 1824 – 1831.

[11] TSUCHIYA H, SHIOTA G. Immune evasion by cancer stem cells [J]. Regenerative therapy, 2021, 17: 20 – 33.

[12] DOGAN F, BIRAY AVCI C. Correlation between telomerase and mTOR pathway in cancer stem cells [J]. Gene, 2018, 641: 235 – 239.

[13] KONG F, ZHENG C, XU D. Telomerase as a "stemness" enzyme [J]. Science China-life science, 2014, 57: 564 – 570.

[14] HUNTLY B J, GILLILAND D G. Leukaemia stem cells and the evolution of cancer-stem-cell research [J]. Nature reviews cancer, 2005, 5: 311 – 321.

[15] SOLTANIAN S, MATIN M M. Cancer stem cells and cancer therapy [J]. Tumour biology, 2011, 32: 425 – 440.

[16] MAENHAUT C, DUMONT J E, ROGER P P, VAN STAVEREN WC. Cancer stem cells: a reality, a myth, a fuzzy concept or a misnomer? An analysis [J]. Carcinogenesis, 2010, 31: 149 – 158.
[17] O'BRIEN C A, POLLETT A, GALLINGER S, et al. A human colon cancer cell capable of initiating tumour growth in immunodeficient mice [J]. Nature, 2007, 445: 106 – 110.
[18] RICCI-VITIANI L, LOMBARDI D G, PILOZZI E, et al. Identification and expansion of human colon-cancer-initiating cells [J]. Nature, 2007, 445: 111 – 115.
[19] KLONISCH T, WIECHEC E, HOMBACH-KLONISCH S, et al. Cancer stem cell markers in common cancers—therapeutic implications [J]. Trends in molecular medicine, 2008, 14: 450 – 460.
[20] ANNABI B, ROJAS-SUTTERLIN S, LAFLAMME C, et al. Tumor environment dictates medulloblastoma cancer stem cell expression and invasive phenotype [J]. Molecular cancer research, 2008, 6: 907 – 916.
[21] SHIEN K, TOYOOKA S, ICHIMURA K, et al. Prognostic impact of cancer stem cell-related markers in non-small cell lung cancer patients treated with induction chemo-radiotherapy [J]. Lung cancer, 2012, 77: 162 – 167.
[22] BERTOLINI G, ROZ L, PEREGO P, et al. Highly tumorigenic lung cancer CD133 + cells display stem-like features and are spared by cisplatin treatment [J]. Proceedings of the National Academy of Science of the United States of America, 2009, 106: 16281 – 16286.
[23] PATRAWALA L, CALHOUN T, SCHNEIDER-BROUSSARD R, et al. Highly purified CD44 + prostate cancer cells from xenograft human tumors are enriched in tumorigenic and metastatic progenitor cells [J]. Oncogene, 2006, 25: 1696 – 1708.
[24] FANG D, NGUYEN T K, LEISHEAR K, et al. A tumorigenic subpopulation with stem cell properties in melanomas [J]. Cancer research, 2005, 65: 9328 – 9337.
[25] KURE S, MATSUDA Y, HAGIO M, et al. Expression of cancer stem cell markers in pancreatic intraepithelial neoplasias and pancreatic ductal adenocarcinomas [J]. International journal of oncology, 2012, 41: 1314 – 1324.
[26] WANG Q, CHEN Z G, DU C Z, et al. Cancer stem cell marker CD133 + tumour cells and clinical outcome in rectal cancer [J]. Histopathology, 2009, 55: 284 – 293.
[27] BOMAN B M, WICHA M S. Cancer stem cells: a step toward the cure [J]. Journal of clinical oncology, 2008, 26: 2795 – 2799.
[28] MANIOTIS A J, FOLBERG R, HESS A, et al. Vascular channel formation by human melanoma cells in vivo and in vitro: vasculogenic mimicry [J]. American journal of pathology, 1999, 155: 739 – 752.
[29] FAN Y L, ZHENG M, TANG Y L, et al. A new perspective of vasculogenic mimic-

ry: EMT and cancer stem cells (Review) [J]. Oncology letters, 2013, 6: 1174-1180.

[30] HANAHAN D, WEINBERG R A. Hallmarks of cancer: the next generation [J]. Cell, 2011, 144: 646-674.

[31] KEITH B, SIMON M C. Hypoxia-inducible factors, stem cells, and cancer [J]. Cell, 2007, 129: 465-472.

Chapter 6 Organoids
第六章 类器官

［中文导读］

类器官（organoid）是一种在体外进行诱导分化形成的，在结构和功能上都类似目标器官或组织的三维细胞复合体。它具有三个特性：能自组装形成，含有多种类型的细胞，在结构和功能上都与体内相应器官有很大程度的相似。目前，人们发现类器官不仅可用于药物的毒性检测、药效评价和新药筛选等，用于建立疾病模型以研究遗传病、传染病和癌症，还可用于精准医疗、研究组织器官发育及用于组织器官的移植和修复。

迄今为止，来源于多种器官的类器官已面世，包括脑、肠道、胃、舌、甲状腺、胸腺、睾丸、肝脏、胰腺、皮肤、肺、肾、心脏及视网膜等。除了来源于健康组织的类器官，大量疾病模型（包括肿瘤模型）的类器官也不断涌现。

当前培养的类器官已初具雏形，因其市场应用前景光明，其发展势头必将迅猛。类器官为药物和疾病的研究提供了新的平台和工具，其能预测药物的治疗反应和效果，为特殊患者的个体化治疗提供依据，为濒死患者带来希望，是人类对干细胞的研究和应用取得的新的认识与突破。类器官作为一个新的研究模型，与传统研究模型相比有很多的优点，但是仍然存在不足。类器官培养的一个重要的内在限制是缺少间充质结构、血管和免疫细胞，因此还须进一步探索与完善。

1 Introduction to Organoids

Starting as a major technological breakthrough, organoids are now well-established as an essential tool in biological research and also have important implications for clinical use. They contain multiple organ-specific cell types which are spatially organized in a manner similar to the organ. In addition, they recapitulate some specific organ functions.

So far, organoids from multiple organs have been created, which include brain, intestine, stomach, tongue, thyroid, thymus, testis, liver, pancreas, skin, lung, kidney, heart, and retina. In addition to the healthy organoids, a plethora of disease models including tumor models, have been developed. Many potential applications of this technology are just beginning to be explored.

1.1 Definition

An organoid is defined as a three-dimensional (3D) *in vitro* grown cell clusters derived from adult stem cells (ASCs) /tissues or pluripotent stem cells, which has multiple organ-specific cell types and is capable of self-renewal and self- organization. Organoid techniques provide unique platforms to model organ development and human diseases. Nowadays, organoids have been widely used in tumor research and scientists have already successfully developed various organoids such as lung, gastric, intestinal, liver, kidney and cerebral organoids.

1.2 Development of Organoids

The appearance of organoid was the product of many fundamental researches in the early twentieth century, such as the study of tissue culture technique, the analysis of dissociated cells, and the study of collagen. Besides, the term "organoid" first appeared in the study of dermoid cysts in 1946[1]. Until the end of twentieth century, organoids revealed their first popularity, mostly in classic developmental biology experiments that sought to describe organogenesis by cell dissociation and reaggregation experiments[2]. In 1975, James Rheinwald and Howard Green described the first long-term culture of normal human cells and they were the first to reconstitute 3D tissue structure from cultured human stem cells. Then, Green and co-workers performed the first successful treatment of two third-degree burn patients with cultured autologous keratinocyte sheets at the Peter Bent Brigham Hospital in 1980. After that, due to the poor understanding of the function of microenvironment that regulate and induce the self-renewal and differentiation of stem cells, as well as the failure of long-term organoids culture *in vitro*, this technology had little progress.

Since Hans Clevers et al. [3] constituted intestinal organoids from cultured intestinal stem cells without a mesenchymal niche in 2009, the past ten years have witnessed a revival of the organoid. Hans Clevers arrived at the breakthrough after a decade of studying the intestinal tract's remarkable ability for self-repair. The researcher suspected intestinal stem cells were key drivers of gut regeneration for a long time. He became the world's first scientist to locate a class of these stem cells, known as Lgr5 stem cells, inside glands (or "crypts") within the intestinal lining. After pioneering a genetic technique to grow Lgr5 stem cells in the laboratory, Clevers was able to develop "intestinoids" that could be used to test drugs safely on active human intestinal tissue.

A few millimetres in size and grown over the course of ten days, these mini-organs have opened the door to grow other "organoids" in the laboratory, including stomach, colon, pancreas and liver organoids. The main area of impact lies in both pharmaceutical development and personalized medicine. The technique also allows replicas of cancerous tumors such as gastric cancer organoids, mammary cancer organoids, colon epithelium organoids, and lung cancer organoids to be grown in order to pretest hundreds of different drug regimens *in vitro*.

1.3 Origin of Organoids

According to the different sources of stem cells, organoids can be divided into two types: organoids derived from stem cells (including ASCs and pluripotent stem cells) and organoids derived from primary tissue.

1.3.1 Adult Stem Cell-Derived Organoids

ASCs are immature precursor cells in organs. They are also the basis for organ cells to reassemble and arrange after *in vitro* culture and play an important role in the process of tissue renewal. However, ASCs can only be differentiated into one or two closely related cell types and can develop and form the similar organs from which they originate. The establishment method of ASCs type organoids: ASCs were extracted from adult organ tissues, and the microenvironment of the organoid development was simulated by using 3D culture technology and adding signal factors, so as to develop and form the ASCs type organoids (Fig. 6 – 1).

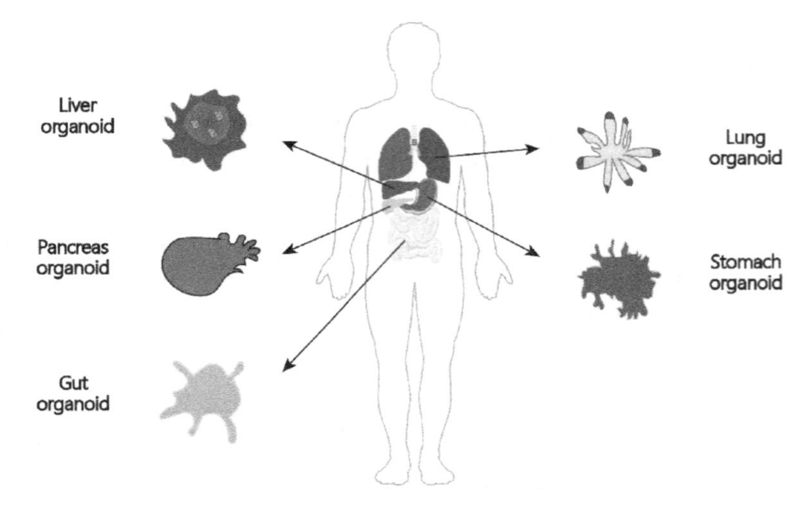

Fig. 6 – 1 Adult stem cells derived organoids

Illustrated by 石培霖.

Reference:

HUCH M, KOO B K. Modeling mouse and human development using organoid cultures [J]. Development, 2015, 142 (18): 3113 – 3125.

ASC is more direct than ESCs and iPSCs, and its application directions are more obvious. It has a great prospect in the direction of physiological function and pathological change mechanism.

1.3.2 Pluripotent Stem Cells-Derived Organoids

ESC is derived from the blastocyst inner cell mass of cells. It has the ability of multilayer differentiation and can proliferate and differentiate infinitely *in vitro*[4]. iPSCs have the same ability of multidirectional differentiation as ESCs[5] (Fig. 6 – 2). The difference is that iP-

SCs are formed by reprogramming the body's cells. The main methods of reprogramming are cell fusion, nuclear transplantation or pluripotent factor overexpression, which stimulate ESCs and iPSCs to differentiate into different substrates by specific signaling pathways[6-7].

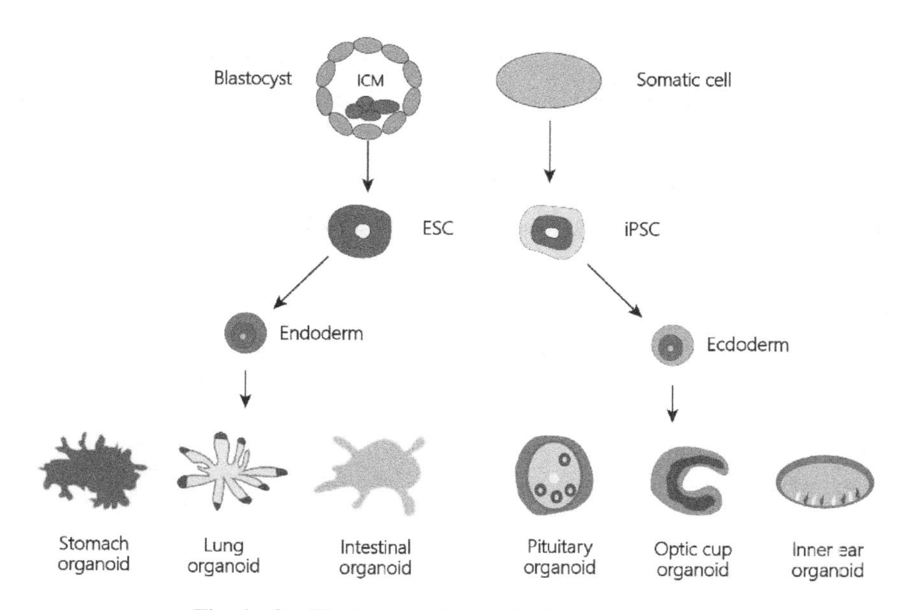

Fig. 6 - 2 Pluripotent stem cells derived organoids

Illustrated by 石培霖.

Reference:

HUCH M , KOO B K. Modeling mouse and human development using organoid cultures [J]. Development, 2015, 142 (18): 3113 - 3125.

1.3.3 Tumor Tissues-Derived Organoids

Tumor cells were cultured into sphere-like colonies by using medium containing special cytokines under 3D culture conditions. The culture products can be cultured and passed down for a long time, showing the same gene expression profile, mutation site, tumor marker, histomorphology, and clinicopathological type as the tumor from which it originated. After transplantation, tumors can be formed in nude mice, and it has similar clinical characteristics with the original tumors, such as invasion and metastasis.

Tumor tissue organoids are cultured from primary or metastatic tumors *in vitro* models. At present, scholars have been successfully developed a variety of tumor tissue organoids, including from the esophagus, stomach, liver, pancreas, colon, breast and endometrial, primary tumor and metastatic lesions of colon, prostate, and breast tumor.

1.4 The Culture of Organoids

The process of organoid culture is as follows.

(1) Obtain stem cells. First of all, harvest tissue under aseptic condition and immerse

it in culture media/tissue preservative solution. Then, wash tissue and mince it into fragments. Finally, wash fragments again and the pure fraction containing stem cells can be used to culture or to dissociate stem cells.

(2) Embed stem cells in Matrigel. At first, create the acellular bottom layer in the inner dish. Next, mix the fragments or stem cells with collagen solution to create the cell-containing top layer. Finally, place the top layer onto the bottom layer and move them in a new outer dish.

(3) Add specific culture medium (containing various niche factors). After the polymerization of Matrigel, specific culture medium is added to the outer dish.

(4) Identify organoids. Use a complete system of organoid identification (such as histological assays, immunohistochemical staining, immunofluorescence, and next-generation sequencing) to check whether the organoids have similarly structural and functional properties to those of the source organs. Fig. 6 – 3 shows the process of organoid culture.

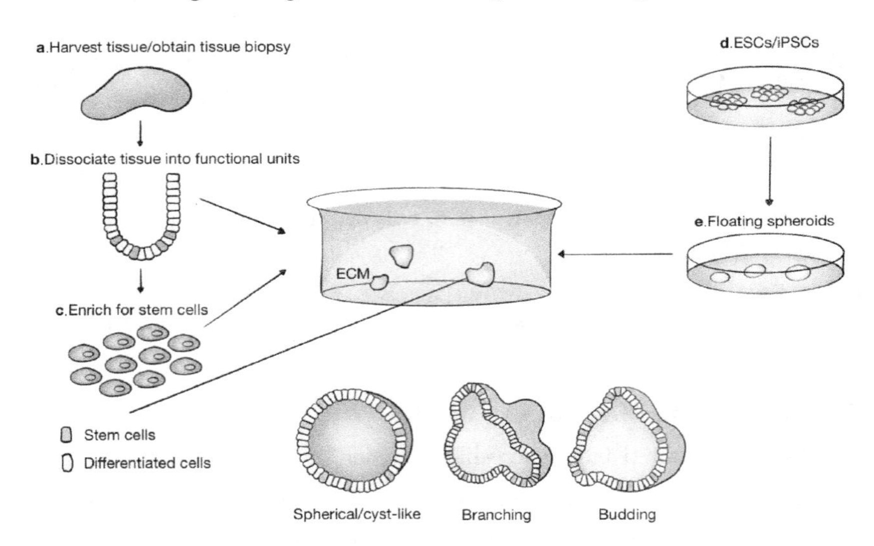

Fig. 6 – 3 The process of organoid culture

Illustrated by 韩雨航.

References:

[1] BARKER N, CLEVERS H. Leucine-rich repeat-containing G-protein-coupled receptors as markers of adult stem cells [J]. Gastroenterology, 2010, 138 (5): 1681 – 1696.

[2] MONTESANO R. Induction of epithelial tubular morphogenesis in vitro by fibroblast-derived soluble factors [J]. Cell, 1991, 66 (4): 697 – 711.

1.5 The Composition of Organoids

1.5.1 Stem Cells

Stem cells are the most important part in the culture of organoid because they have the ability of both self-renewal and differentiation, which means a single stem cell can replicate itself or differentiate into many cell types. Stem cells are responsible for the growth, homeostasis and repair of many tissues. In human body, stem cells can be found in the developing embryo, cord blood, various adult tissues, and tumors.

For example, in the human body, the epithelium of the small intestine has a higher self-renewal rate than any other tissue, with a turnover time of less than five days. Intestinal stem cells reside the bottom of the intestinal crypt. They divide, amplify and flow onto the vill, where they differentiate, absorb nutrients and eventually die at the villus tips, so the organoid culture can use not only the fragments of pure crypts, but also the stem cells isolated from the pure crypts. The intestinal stem cells can differentiate into many cell types including absorptive enterocytes, multiple secretory cells (paneth cells, goblet cells, enteroendocrine cells, and tuft cells), and the M cells that cover Peyer's patches.

1.5.2 Niche Factors

Stem cell niche harbors stem cells and provides external control of stem cells. The niche contains many niche factors that are important to maintain the function of stem cells such as self-renewal, proliferation and differentiation. The aim of adding niche factors to medium is to create an environment *in vitro* which is similar to the niche *in vivo*.

For example, intestinal stem cells are cultured in specific medium supplemented with three niche factors: EGF, R-spondin-1, and Noggin. First of all, Wnt signalling is a pivotal requirement for crypt proliferation and the Wnt agonist R-spondin-1 induces marked crypt hyperplasia *in vivo*. Next, epidermal growth factor (EGF) signals exert strong mitogenic effect on stem cells and transit amplifying cells (TA cells). Last but not least, transgenic expression of Noggin induces an expansion of crypt numbers.

The Tab. 6 – 1 shows different niche factors used in the culture of different organoids.

Tab. 6 – 1 Growth factors used in different organoids

Types of organoids	Cultural conditions	References
Mammary cancer organoids	FGF7/10, EGF, R-spondin, Noggin	[8]
Prostate cancer organoids	EGF, Noggin, R-spondin-1, FGF10, FGF2	[9]
Gastric cancer organoids	EGF, Noggin, R-spondin-1, Wnt, FGF10	[10]
Liver cancer organoids	FGF10, EGF, HGF	[11]
Endometrium cancer organoids	EGF, Noggin, R-spondin-1, FGF10, HGF	[12]

1.5.3 Matrigel

The former research observed that the floatation of the gels has created a more permissive environment for differentiation, so Richard Swarm and his group isolated a gel with characteristics of the basement membrane and named it as "EHS" sarcoma using the initials of the three investigators—Engelbreth, Holm, and Swarm. This was the discovery of what we know today as Matrigel.

The function of Matrigel is similar to basement membrane *in vivo*. Basement membrane provides structural support for stem cells, so does Matrigel. Stem cells are embedded and suspended in Matrigel, and Matrigel can provide a 3D zone for proliferation and differentiation of stem cells and replace feeder cells in traditional culture systems[13-14].

Due to the discovery of the components of basement membrane and matrix, Matrigel today is consisted of collagen, laminin, entactin, proteoglycan, and some growth factors. These components can promote the expansion of stem cells and anchor the structure of cultured tissues.

Take the culture of small intestinal organoids as an example. As shown in Fig. 6-4, the first step is to obtain small intestinal stem cells. To begin with, crypt isolation. Crypts were released from murine small intestine and washed with cold PBS. The tissue was then chopped into fragments and further washed. After incubation, the villous fraction must be removed. Isolated crypts were centrifuged to separate crypts from single cells. The final fraction consisted of essentially pure crypts and was used for culture or single cell dissociation. Next, stem cell dissociation. Encouraged by the observation that Lgr5-positive intestinal stem cells can be long-term alive and proliferate actively *in vivo* as well as Lgr5-positive cells constitute multipotent stem cells that generate all cell types of the epithelium. Small intestinal stem cells can be isolated by the marker gene Lgr5 (leucine-rich repeat containing G-protein-coupled receptor 5)[15].

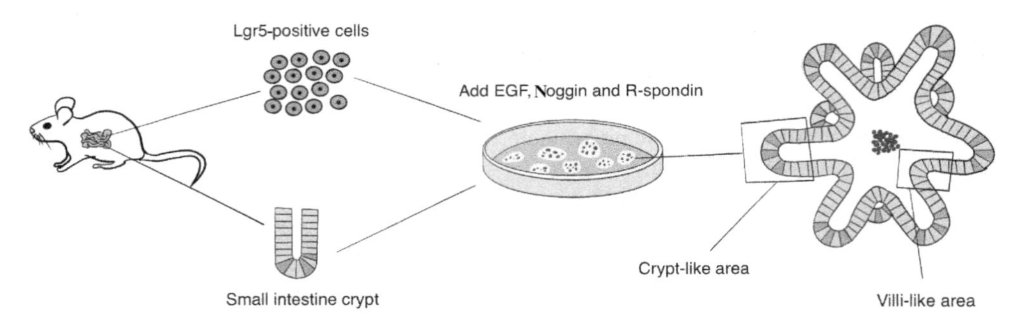

Fig. 6-4 The small intestine organoid culture process

Illustrated by 韩雨航.

References:

[1] SATO T, VRIES R G, SNIPPERT H J, et al. Single Lgr5 stem cells build crypt-villus structures *in vitro* without a mesenchymal niche [J]. Nature, 2009, 459, 262-265.

[2] BARKER N, CLEVERS H. leucine-rich repeat-containing g-protein-coupled receptors as markers of adult stem cells [J]. Gastroenterology, 2010, 138 (5), 1681-1696.

The second step, because laminin (α1 and α2) is enriched at the crypt base, the researcher used the laminin-rich and collagen-rich Matrigel, which can simulate the growing environment of intestinal cells. Meanwhile, the pure crypts or single Lgr5-positive stem cells are embedded in Matrigel and cultured in serum-free medium supplemented with three recombinant proteins: R-spondin-1 (a Wnt signal amplifier and ligand of Lgr5), EGF, and the BMP inhibitor Noggin.

Finally, multiple methods such as microarray analysis and electron microscopic analysis revealed that organoids remained highly similar to freshly isolated small intestinal crypts.

1.6 Identification of Organoids

In order to evaluate the characteristics and physiological functions of organoids, it is often necessary to identify the morphological characteristics of organoid cells and markers related to development and differentiation. Now a complete system of organoid identification has been established, such as histological assays, immunohistochemical staining, immunofluorescence, next-generation sequencing (NGS), etc.

Histological evaluation and immunohistochemistry are used to investigate whether patient-derived organoids (PDOs) and parental biopsies show notable morphological similarities. For example, we first establish a culture system of organoids *in vitro*. After growing for several days, we prepare them into paraffin sections, then detected by immunohistochemistry to verify parental tumor's expression pattern is maintained in PDOs.

Based on the growth of organoids at the 3D level, the immunofluorescence technique applied at the 3D level was explored. Such 3D signal detection can conveniently and comprehensively present the proliferation and differentiation of organs on the basis of maintaining the 3D structure of the sample, which is conducive to master the dynamic process of the continuous growth of organoids.

1.7 The Comparison of Organoids with Other Models

Traditional *in vitro* models usually refer to cell lines and animal models. Cell lines are 2D cell cultures, which have limitations in fully representing cancer structures, size, density, and more importantly, tumor heterogeneity in a 3D perspective. These drawbacks make the results of cell lines in tumor studies and preclinical testing inaccurate from their original clinical traits. Animal models are physiologic but experimentally complex, and some aspects of human development and disease are not easily or accurately modeled in animals. Recently, researchers have developed better preclinical models by using patient-derived tumor xenografts (PDTXs), which can accurately demonstrate cancer biology because they are patient-derived culture systems. Among the three mentioned culture systems, they can efficiently model various cancer phenotypes for tumorigenesis and metastasis, drug responses, and drug resistances. However, they are expensive, time-consuming and lacking immune responses.

Organoids are the intermediates of the three models. They can recapitulate sophisticated tumor microenvironments and heterogeneity fairly accurately. Moreover, they are highly sensitive to high-throughput drug screening than PDTXs and have a high initiation efficiency. They are inexpensive and faster than PDTXs. Compared with cell lines and animal model, organoid shows great advantages with its higher recapitulate fitting degree, shorter culture cycle, more stable quality during passaging and etc. Unlike traditional *in vitro* 2D cultures, organoids are similar to original tissues in their genetic composition and cellular architecture, harboring small populations of genetically stable, self-renewing stem-like cells that can give rise to fully differentiated progeny, comprising major cell lineages at frequencies similar to those in living tissue.

The Tab. 6 – 2 shows comparison of organoid with other models.

Tab. 6 – 2 Comparison of organoid with other models

Items	2D cell culture	Organoid	Animal model
Physiologic representation	Limited	Semiphysiologic	Physiologic
Vascularization and immune system	No	No	Yes
High throughput screening	Yes	Yes	No
Manipulability	Excellent	Good, but have many experimental variability	Limited
Biobanking	Yes	Yes	Yes, but only at the cellular level
Genome editing	Yes	Yes	Yes, but may require generation of ESCs
Modeling organogenesis	Poor	Yes, reduced complexity, suitable for study of cell-cell communication, morphogenesis	Yes, but often confounded by complex tissue environment
Modeling human development and disease	Poor	Yes	Yes

In summary, organoids can compensate for the drawbacks mentioned in other models, while not without disadvantages. Firstly, present organoids cannot maintain their vasculature, immune responses and peripheral nerve systems. Secondly, ethics problems, including the use of human embryos, need to be considered and dealt with carefully. Thirdly, they lack specific cultivating media. Therefore, they still need improvements.

2 Applications of Organoids

2.1 Cancer Medicine

Patient-derived organoids are 3D tissue cell clusters derived from tissues or tumor specific stem cells, which can recapitulate characteristics of tumors and heterogeneity of tumor cells *in vivo*. Tumor organoids retain the molecular and cellular composition of the original tumor, can almost accurately recapitulate tumor heterogeneity and microenvironment, therefore, provide a tool to verify the mechanism of tumor formation and develop personalized therapies. Tumor organoids can more accurately predict the patient's treatment response, conduct drug screening, and develop a personalized treatment plan for the patient (Fig. 6 – 5). Tumor organoids are also used to study the fundamental issues of tumorigenesis and metastasis.

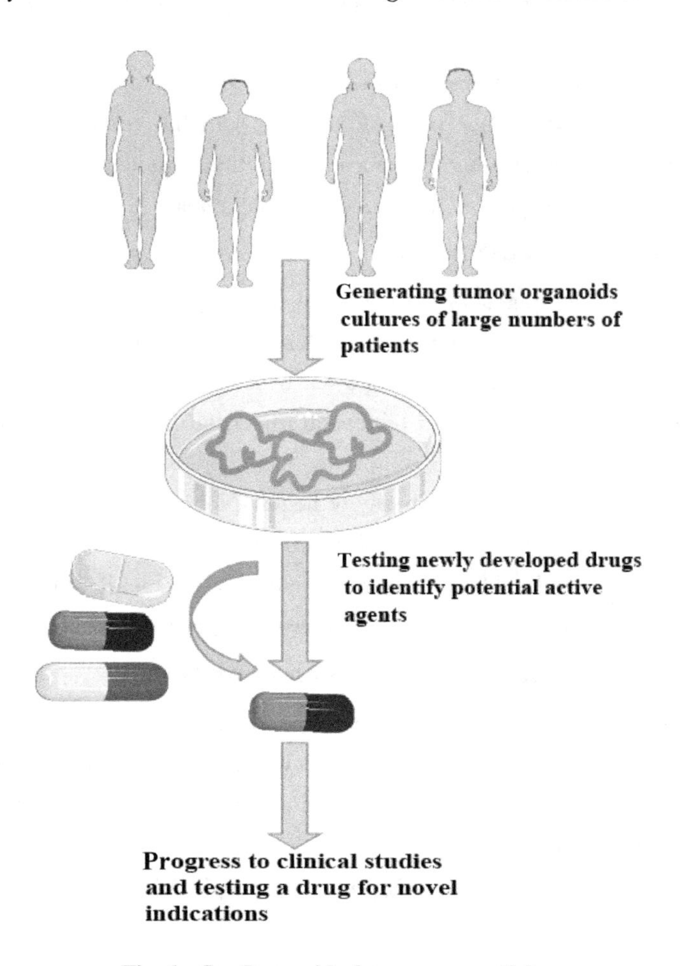

Fig. 6 – 5 Organoids for cancer medicine

Illustrated by 黄如凡.

Reference:

WEEBER F, OOFT S N, DIJKSTRA K K, et al. tumor organoids as a pre-clinical cancer model for drug discovery [J]. Cell chemical biology, 2017, 24 (9): 1092 – 1100.

More importantly, *in vitro* tumor organoid cultures can potentially be used to assess chemosensitivity to diverse antineoplastic agents. For instance, D Fiore et al. used organoid of primary colon cancer stem cells to demonstrate that rimonabant was able to reduces both tumor differentiated cells and colon cancer stem cells proliferation and to control their survival in long term cultures[16].

2.2 Precision Medicine

Organoids can better recapitulate *in vivo* characteristics in phenotype, genotype, and specific functions as well as physiological and pathological changes even after many generations. They are endowed with enormous potential to identify the feasible optimized treatment strategy for the individual patient.

Researchers have compared responses to anticancer agents *in vivo* in organoids and PDOs-based orthotopic mouse tumor xenograft models with the responses of the patients in clinical trials, the data suggests that PDOs can recapitulate patient responses in the clinic and could be implemented in personalized medicine programs[17].

2.3 Regenerative Medicine

The purpose of modern regenerative medicine is to replace disease tissue with a corresponding healthy tissue through allogeneic transplantation. Organoids can be used to replace donor tissue. For example, M Tamai constructed a liver organoid tissue that could be comparable to the endogenous liver by using a bioreactor, may fulfill the fundamental requirements of a donor tissue[18].

However, in addition to the short supply of healthy donor tissue, internal immune rejection poses a serious threat to the long-term survival and function in the recipient's body. By using patient biopsies or easily obtained tissues to cultivate organoids of the same genotype or match the patient's lymphocyte antigens, thereby avoiding immune rejection.

2.4 Drug Development: Screening and Toxicology

Current preclinical models fail to assess the toxicity and side effects of drugs accurately, which lead to failures of new drugs research and development (R&D). Organoids are highly similar to physiological tissues and can be used to simulate the experimental drug response, and normal organoids can also be used to assess drug toxicity. Drug-related hepatotoxicity is inspiringly observed in hepatic organoids at near-physiological levels. Cardiac adverse effects such as arrhythmias and cardiotoxic effects can also be tested in 3D cultures. Besides, organoid can be used to simulate pharmacokinetics of new drugs.

Many organoids can undergo expand extensively in culture and maintain genome stability, which makes them suitable for high-throughput screens (HTPS) of drugs. PDOs can be used for high-throughput drug screening to test drug chemosensitivity for personalized therapy. For example, S. N. Ooft et al. used PDO test predicted response of the biopsied lesion in more than 80% of patients treated with irinotecan-based therapies without misclassifying patients who would have benefited from treatment, the data suggest that PDOs could be used to predict response to chemotherapy in metastatic colorectal cancer patients[19].

2.5 Disease Modeling

Many diseases lack vitro model, but normal human epithelium can be cultured by the technology of organoids to provide models for the study of pathogenic mechanism of disease. In recent years, organoids are crucial tools for disease modeling, which have been used as disease models in a variety of diseases, including infectious diseases, genetic diseases, neurodegenerative diseases, etc.

McCracken et al. report the de novo generation of 3D human gastric tissue *in vitro* through the directed differentiation of human pluripotent stem cells. Using human gastric organoids to model pathogenesis of human disease, they found that H. pylori infection resulted in rapid association of the virulence factor CagA with the c-Met receptor, activation of signaling and induction of epithelial proliferation[20].

2.6 Basic Research

Organoids have great potential in basic research in biomedicine. This emerging tool enables people to study organ formation, lineage differentiation, and dynamic balance within tissues *in vitro*. The human cerebral cortex possesses structural and functional features that are not found in the lower species traditionally used to model brain development and disease. Accordingly, scientists have been placing attention on directing pluripotent stem cells to form human brain-like structures termed organoids.

Nowadays, organoids are widely used in modeling brain development. For instance, to investigate how ZIKV infection leads to microcephaly, Dang J used hESCs derived cerebral organoids to recapitulate first trimester fetal brain development[21].

3 Limitation of Organoids

Currently, cultured organoids are simple in structure, but only have the characteristics of embryonic development stage of their corresponding organs. It does not contain all types of cellular and structural features of the organ, which will inevitably affect its authenticity and reliability in pharmaceutical research and clinical application. Organoids can only partially replace animal models of disease and provide some basis and reference.

As a new research model, organoids have many advantages, but still face some problems that need to be solved. On the one hand, compared with living cells, organoids lack many cellular components that are contained within an *in vivo* system (such as stromal, vascular endothelial, or immune cells). The biology and behavior of cancer can be influenced by microenvironmental factors including the cell types, which can explain differences in drug sensitivity for the same cancer cells grown in 2D or 3D cultures or monoculture or coculture with fibroblasts[22-23]. This does not fully represent the environment in which tumors are located in the body. On the other hand, in the process of organoid culture, it is necessary to add growth factors and various chemical reagents in the culture medium, which may cause gene mutation and abnormal growth of cells. The external deficiency of organoid culture lies in the substitutes of ECM, such as matrix gel, basement membrane extract, or fetal bovine serum containing some uncertain components which may affect the results of drug experiments. Therefore, it is necessary to establish a 3D culture system model of multiple cell types (such as fibroblasts, immune cells, and endothelial cells) to reflect the effects of ECM, epithelial matrix communication, cell matrix interaction, and cell-cell crosstalk[24]. This reduces the impact of the ECM.

Supplement

List of Abbreviations	
ASCs	adult stem cells
EGF	epidermal growth factor
ESCs	embryonic stem cells
FGF	fibroblast growth factor
HGF	hepatocyte growth factor
HTPS	high-throughput screens
iPSCs	induced pluripotent stem cells
NGS	next-generation sequencing
PDOs	patient-derived organoids
PDTXs	patient-derived tumor xenografts

Key Words List	
高通量筛选	high-throughput screens
基因表达谱	gene expression profile
基因测序	gene sequencing
基质胶	matrigel
类器官	organoid
免疫荧光	immunofluorescence
免疫组化染色	immunohistochemical staining
人源肿瘤异种移植模型	patient-derived tumor xenografts
胎牛血清	fetal bovine serum
有丝分裂	mitosis

(To be continued)

Key Words List	
肿瘤发生	tumorigenesis
肿瘤类器官	patient-derived organoids
组织形态学	histomorphology

References

[1] SMITH E, COCHRANE W J. Cystic organoid teratoma: (report of a case) [J]. Canadian medical association journal, 1946, 55: 151 – 152.

[2] CLEVERS H. Modeling development and disease with organoids [J]. Cell, 2016, 165: 1586 – 1597.

[3] SATO T, VRIES R G, SNIPPERT H J, et al. Single Lgr5 stem cells build crypt-villus structures in vitro without a mesenchymal niche [J]. Nature, 2009, 459: 262 – 265.

[4] BARKER N, CLEVERS H. Leucine-rich repeat-containing G-protein-coupled receptors as markers of adult stem cells [J]. Gastroenterology, 2010, 138: 1681 – 1696.

[5] SATO T, STANGE D E, FERRANTE M, et al. Long-term expansion of epithelial organoids from human colon, adenoma, adenocarcinoma, and Barrett's epithelium [J]. Gastroenterology, 2011, 141: 1762 – 1772.

[6] HUCH M, BONFANTI P, BOJ S F, et al. Unlimited in vitro expansion of adult bi-potent pancreas progenitors through the Lgr5/R-spondin axis [J]. EMBO journal, 2013, 32: 2708 – 2721.

[7] BARTFELD S, BAYRAM T, VAN DE WETERING M, et al. *In vitro* expansion of human gastric epithelial stem cells and their responses to bacterial infection [J]. Gastroenterology, 2015, 148: 126 – 136. e6.

[8] SACHS N, DE LIGT J, KOPPER O, et al. A Living biobank of breast cancer organoids captures disease heterogeneity [J]. Cell, 2018, 172: 373 – 386. e10.

[9] SUGIMOTO S, OHTA Y, FUJII M, et al. Reconstruction of the human colon epithelium *in vivo* [J]. Cell stem cell, 2018, 22: 171 – 176. e5.

[10] KIM M, MUN H, SUNG C O, et al. Patient-derived lung cancer organoids as *in vitro* cancer models for therapeutic screening [J]. Nature communication, 2019, 10: 3991.

[11] THOMSON J A, ITSKOVITZ-ELDOR J, SHAPIRO S S, et al. Embryonic stem cell lines derived from human blastocysts [J]. Science, 1998, 282: 1145 – 1147.

[12] TAKAHASHI K, YAMANAKA S. Induction of pluripotent stem cells from mouse embryonic and adult fibroblast cultures by defined factors [J]. Cell, 2006, 126: 663 – 676.

[13] BLAU H M, CHIU C P, WEBSTER C. Cytoplasmic activation of human nuclear genes in stable heterocaryons [J]. Cell, 1983, 32: 1171 – 1180.

[14] CHUNG Y G, EUM J H, LEE J E, et al. Human somatic cell nuclear transfer using adult cells [J]. Cell stem cell, 2014, 14: 777 – 780.

[15] SACHS N, DE LIGT J, KOPPER O, et al. A living biobank of breast cancer organoids captures disease heterogeneity [J]. Cell, 2018, 172: 373 – 386. e10.

[16] GAO D, VELA I, SBONER A, et al. Organoid cultures derived from patients with advanced prostate cancer [J]. Cell, 2014, 159: 176 – 187.

[17] BARTFELD S, BAYRAM T, VAN DE WETERING M, Et al. *In vitro* expansion of human gastric epithelial stem cells and their responses to bacterial infection [J]. Gastroenterology, 2015, 148: 126 – 136, e6.

[18] BROUTIER L, MASTROGIOVANNI G, VERSTEGEN M M, et al. Human primary liver cancer-derived organoid cultures for disease modeling and drug screening [J]. Nature medicine, 2017, 23: 1424 – 1435.

[19] TURCO M Y, GARDNER L, HUGHES J, et al. Long-term, hormone responsive organoid cultures of human endometrium in a chemically defined medium [J]. Nature cell biology, 2017, 19: 568 – 577.

[20] FUJII M, SHIMOKAWA M, DATE S, et al. A Colorectal tumor organoid library demonstrates progressive loss of niche factor requirements during tumorigenesis [J]. Cell stem cell, 2016, 18: 827 – 838

[21] MONTESANO R, SCHALLER G, ORCI L. Induction of epithelial tubular morphogenesis in vitro by fibroblast-derived soluble factors [J]. Cell, 1991, 66: 697 – 711.

[22] XU C, INOKUMA M S, DENHAM J, et al. Feeder-free growth of undifferentiated human embryonic stem cells [J]. Nature biotechnology, 2001, 19: 971 – 974.

[23] FIORE D, RAMESH P, PROTO M C, et al. Rimonabant kills colon cancer stem cells without inducing toxicity in normal colon organoids [J]. Frontiers in pharmacology, 2018, 8: 949.

[24] VLACHOGIANNIS G, HEDAYAT S, VATSIOU A, et al. Patient-derived organoids model treatment response of metastatic gastrointestinal cancers [J]. Science, 2018, 359: 920 – 926.

Chapter 7 Different Types of Organoids
第七章 不同类型的类器官

［中文导读］

类器官的研究和发展是过去十数年里干细胞研究领域的关键进展之一。在过去的十数年里，研究人员从其他领域（如生物工程）中汲取灵感，并借鉴多种生物医学技术，以培养出更具生理相关性和更适于实际应用的类器官。迄今为止，已有多种健康组织或疾病模型来源的类器官培养被报道。鉴于类器官在生理基础研究、药物开发和再生医学等领域中的重要意义和广泛应用，我们期待更多适于基础研究和实际应用的类器官被开发和培养出来。

1 Different Types of Organoids

In the past decade or so, the field of organoids has developed rapidly, and more researches on organoids from different sources have been reported, including healthy organs/tissues and tumor-derived organoids, which provides more possibilities for drug and disease research. Tab. 7 – 1 lists the main different types of organoid researches published in the last five years.

Tab. 7 – 1 The lists of different types of organoid researches published in the last five years

Cancer	Year	Reference
Liver cancer organoid	2017	BROUTIER L, MASTROGIOVANNI G, VERSTEGEN M M, et al. Human primary liver cancer-derived organoid cultures for disease modeling and drug screening [J]. Nature medicine, 2017, 23: 1424 – 1435.
	2018	NUCIFORO S, FOFANA I, MATTER M S, et al. Organoid models of human liver cancers derived from tumor needle biopsies [J]. Cell reports, 2018, 24: 1363 – 1376.
	2019	SUN L, WANG Y, CEN J, et al. Modelling liver cancer initiation with organoids derived from directly reprogrammed human hepatocytes [J]. Nature cell biology, 2019, 21: 1015 – 1026.
	2019	LI L, KNUTSDOTTIR H, HUI K, et al. Human primary liver cancer organoids reveal intratumor and interpatient drug response heterogeneity [J]. JCI Insight, 2019, 4: e121490.

(To be continued)

Cancer	Year	Reference
Colorectal cancer organoid	2009	SATO T, VRIES R G, SNIPPERT H J, et al. Single Lgr5 stem cells build crypt-villus structures in vitro without a mesenchymal niche [J]. Nature, 2009, 459: 262 - 265.
	2011	SATO T, STANGE D E, FERRANTE M, et al. Long-term expansion of epithelial organoids from human colon, adenoma, adenocarcinoma, and Barrett's epithelium [J]. Gastroenterology, 2011, 141: 1762 - 1772.
	2015	VAN DE WETERING M, FRANCIES H E, FRANCIS J M, et al. Prospective derivation of a living organoid biobank of colorectal cancer patients [J]. Cell, 2015, 161: 933 - 945.
	2016	FUJII M, SHIMOKAWA M, DATE S, et al. A colorectal tumor organoid library demonstrates progressive loss of niche factor requirements during tumorigenesis [J]. Cell stem cell, 2016, 18: 827 - 838.
	2017	FUMAGALLI A, DROST J, SUIJKERBUIJK S J, et al. Genetic dissection of colorectal cancer progression by orthotopic transplantation of engineered cancer organoids [J]. Proceedings of the National Academy of Science of the United States of America, 2017, 114: E2357 - E2364.
	2018	ROERINK S F, SASAKI N, LEE-SIX H, et al. Intra-tumour diversification in colorectal cancer at the single-cell level [J]. Nature, 2018, 556: 457 - 462.
Breast cancer organoid	2007	LEE G Y, KENNY P A, LEE E H, et al. Three-dimensional culture models of normal and malignant breast epithelial cells [J]. Nature methods, 2007, 4: 359 - 365.
	2018	SACHS N, DE LIGT J, KOPPER O, et al. A living biobank of breast cancer organoids captures disease heterogeneity [J]. Cell, 2018, 172: 373 - 386.
	2019	ROELOFS C, HOLLANDE F, REDVERS R, et al. Breast tumour organoids: promising models for the genomic and functional characterisation of breast cancer [J]. Biochemical society transactions, 2019, 47: 109 - 117.
Pancreatic cancer organoid	2015	BOJ S F, HWANG C I, BAKER L A, et al. Organoid models of human and mouse ductal pancreatic cancer [J]. Cell, 2015, 160: 324 - 338.
	2018	SEINO T, KAWASAKI S, SHIMOKAWA M, et al. Human pancreatic tumor organoids reveal loss of stem cell niche factor dependence during disease progression [J]. Cell stem cell, 2018, 22: 454 - 467.

(To be continued)

Cancer	Year	Reference
Prostate cancer organoid	2014	GAO D, VELA I, SBONER A, et al. Organoid cultures derived from patients with advanced prostate cancer [J]. Cell, 2014, 159: 176 – 187.
	2014	KARTHAUS W R, IAQUINTA P J, DROST J, et al. Identification of multipotent luminal progenitor cells in human prostate organoid cultures [J]. Cell, 2014, 159: 163 – 175.
	2014	CHUA C W, SHIBATA M, LEI M, et al. Single luminal epithelial progenitors can generate prostate organoids in culture [J]. Nature cell biology, 2014, 16: 951 – 4.
Ovarian cancer organoid	2018	HILL S J, DECKER B, ROBERTS E A, et al. Prediction of DNA repair inhibitor response in short-term patient-derived ovarian cancer organoids [J]. Cancer discovery, 2018, 8: 1404 – 1421.
	2019	KOPPER O, DE WITTE CJ, LÕHMUSSAAR K, et al. An organoid platform for ovarian cancer captures intra- and interpatient heterogeneity [J]. Nature medicine, 2019, 25: 838 – 849.
Cerebral organoid	2016	QIAN X, NGUYEN H N, SONG M M, et al. Brain-region-specific organoids using mini-bioreactors for modeling ZIKV exposure [J]. Cell, 2016, 165: 1238 – 1254.
	2018	BIAN S, REPIC M, GUO Z, et al. Genetically engineered cerebral organoids model brain tumor formation [J]. Nature methods, 2018, 15: 631 – 639.
	2019	LINKOUS A, BALAMATSIAS D, SNUDERL M, et al. Modeling patient-derived glioblastoma with cerebral organoids [J]. Cell reports, 2019, 26: 3203 – 3211.
Tumor immune	2018	NEAL J T, LI X, ZHU J, et al. Organoid modeling of the tumor immune microenvironment [J]. Cell, 2018, 175: 1972 – 1988.
	2020	YUKI K, CHENG N, NAKANO M, et al. Organoid models of tumor immunology [J]. Trends in immunology, 2020, 41: 652 – 664.
Fallopian tube organoid	2015	KESSLER M, HOFFMANN K, BRINKMANN V, et al. The Notch and Wnt pathways regulate stemness and differentiation in human fallopian tube organoids [J]. Nature communication, 2015, 6: 8989.
Urinary system	2019	SCHUTGENS F, ROOKMAAKER M B, MARGARITIS T, et al. Tubuloids derived from human adult kidney and urine for personalized disease modeling [J]. Nature biotechnology, 2019, 37: 303 – 313.

(To be continued)

Cancer	Year	Reference
Respiratory system	2017	CHEN Y W, HUANG S X, DE CARVALHO ALRT, et al. A three-dimensional model of human lung development and disease from pluripotent stem cells [J]. Nature cell biology, 2017, 19: 542-549.
	2019	SACHS N, PAPASPYROPOULOS A, ZOMER-VAN OMMEN D D, et al. Long-term expanding human airway organoids for disease modeling [J]. EMBO journal, 2019, 38: e100300.

2. Patient-Derived Human Liver Cancers Organoids

2.1 Introduction

Liver cancer is one of the most common malignant tumours of the digestive system and its incidence and mortality rate are increasing year by year. As a country with a high incidence of liver cancer, the mortality rate of liver cancer in China ranks second among all cancers. There are two types of liver cancer, including hepatocellular carcinoma (HCC) and intrahepatic cholangiocellular carcinoma (CCC).

HCC is the most common primary liver cancer. The main risk factors causing HCC include infection with hepatitis B virus, hepatitis C virus, alcoholic liver disease, nonalcoholic fatty liver disease, and nonalcoholic steatohepatitis. CCC is the second most common primary liver cancer with main risk factors including primary sclerosing cholangitis, cysts of the biliary duct, and parasitic infection with liver flukes.

Surgical treatment remains a major choice for early HCC. However, the postoperative 5-year recurrence rate is over 70%, due to the following reasons: high invasiveness, irregularity along the peripheral tissue and blood vessel infiltration, unclear intraoperative boundary difficult to accurately identify, and small satellite (maximum diameter of 5 mm or less) difficult to preoperative or intraoperative imaging diagnosis and effective which cause residual during surgery.

However, most patients with liver cancer are diagnosed at an advanced stage, which is too late to perform an operation. In the past, conventional chemotherapies have been extensively applied to the treatment of advanced liver cancer, but none of them have improved survival. In 2008, the major progress came with the introduction of the multikinase inhibitor sorafenib. Sorafenib is the major treatment option for advanced HCC which can significantly prolong median survival from 7.9 months in the control group to 10.7 months in the sorafenib treatment group. Regrettably, many liver cancer patients develop resistance to sorafenib after taking it for a period of time. Hence, there is an urgent need for new therapies for HCC. A major obstacle in preclinical drug development is the lack of appropriate cell culture model systems. Traditional models usually refer to cell lines and animal models. Cell lines are 2D cell cultures, which have limitations in fully representing cancer structures, size, density, and more importantly, tumor heterogeneity in a 3D perspective. These drawbacks make the results of cell lines in tumor studies and preclinical testing inaccurate from their original clinical traits. Animal models are physiologic but experimentally complex, and some aspects of human development and disease are not easily or accurately modeled in animals. Therefore, the researchers turned to organoid which shows great advantages with its higher recapitulate fitting degree, shorter culture cycle, more stable quality during passaging, etc.

In this article, the authors successfully generate tumor organoid cultures from needle biopsies obtained from HCC patients, and proved that the established HCC organoids recapitulate the histopathological morphological and genetic features of the originating tumor *in vitro*.

2.2 The Framework of the Article

(1) Obtain tumor and non-tumor liver samples from liver cancer patients by ultrasound-guided needle biopsy.

(2) Establish organoid cultures from needle biopsies of HCC and paired non-tumor liver tissues.

(3) Identification of established organoids. Histopathological analysis to investigate whether the histological characteristics of the originating tumors were preserved in the HCC organoids. The H&E staining results showed that the organoids maintained the growth pattern and differentiation grade of the original tumor. The immunohistochemical analysis revealed consistent distribution and expression intensity of AFP (a tumor marker of HCC) between HCC organoids and their original tumor biopsy tissue. Morphological analysis of HCC organoids, tumor organoids formed compact spheroids, whereas liver organoids from non-tumor liver tissue grew as cystic structures. Genetic analysis of HCC organoids to assess whether the HCC organoids recapitulate the genetic alterations of the originating tumor.

(4) Xenografts by injecting HCC organoids subcutaneously into immunodeficient mice, to assess whether HCC organoids retained the ability to form bona fide tumors in mice.

(5) Drug sensitive test to sorafenib. In order to assess whether HCC-derived organoids would be a suitable system for preclinical drug development, the author treated HCC organoid cultures with different concentrations of sorafenib and monitored cell viability.

(6) In conclusion, organoid models can be derived from needle biopsies of liver cancers and provide a tool for developing tailored therapies.

References

NUCIFORO S, FOFANA I, MATTER M S, et al. Organoid models of human liver cancers derived from tumor needle biopsies [J]. Cell reports, 2018, 24: 1363 - 1376.

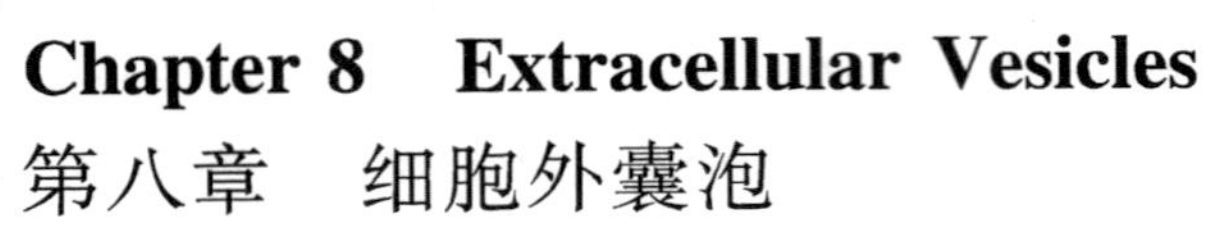

Chapter 8　Extracellular Vesicles
第八章　细胞外囊泡

［**中文导读**］

细胞间通讯是多细胞生物的重要特征，它可以通过细胞间的直接接触或转移分泌的分子来实现。在过去的几十年里，一种新颖的细胞间通讯机制正逐渐被深入研究，也就是大多数细胞释放的脂质膜包裹的纳米级小泡——细胞外囊泡。细胞外囊泡被认为是细胞间通讯的有效载体。细胞外囊泡最初被发现与血小板功能有关，故被称为血小板尘埃，随后有研究者发现网织红细胞中的多泡体将这种囊泡释放到细胞外的空间，于是将其命名为外泌体。在这几十年的研究中，科学家发现几乎所有的哺乳动物细胞都能够纯化出细胞外囊泡，包括干细胞、免疫细胞、神经细胞，以及大量的癌细胞；除此之外，细胞外囊泡还广泛分布于血清、尿液、唾液及其他生物液体中。

细胞外囊泡是一种具有脂质膜的内源性异质纳米囊泡，携带着来源于细胞的生物活性分子，如蛋白质、脂质、核酸，具有高生物相容性和高靶向特异性，能够穿越生物屏障，进入血液、尿液、脑脊液等体液中，到达相邻和远处的靶细胞。细胞外囊泡能参与原核生物和高等真核生物细胞间的生物信号传输，从而影响受体和母细胞的各种生理和病理功能，调节多种多样的生物功能。细胞外囊泡在免疫疾病、癌症、感染性疾病和神经退行性疾病等疾病中的病理生理作用已开始被认识，它能参与免疫调节、促进组织再生，并作为干细胞治疗的潜在替代品。因此，细胞外囊泡正在逐渐被开发成疾病治疗的潜在新靶点。此外，天然的和基因工程修饰的细胞外囊泡具有低免疫原性，因此，它可作为药物载体应用于药物传递。相信在未来，细胞外囊泡可作为诊断、预后和预测性的生物标记物，在临床上发挥治疗各种疾病的潜力。

1　Introduction

Intercellular communication, which involves direct cell-to-cell contact with neighboring cells or transfer of secreted molecules communicate with distant cells, is pivotal for multicellular organisms. In recent years, a new signaling paradigm for intercellular communication involving intercellular exchange of extracellular vesicles (EVs) has become a focus of scientific research. Perhaps unsurprisingly, the initial observations of EVs were regarded as membrane debris without biological property or communication function. Originally, EVs were described as procoagulant platelet-derived particles or platelet dust in plasma[1-2]. Subsequently, EVs were identified as matrix vesicles that played a role during bone calcification[3]. The term exosome was originally used to name exfoliated vesicles with 5'-nucleotidase activity which derived from a variety of cultured cells and ranged from 40 to 1 000 nm, and supposed may serve a physiologic function[4]. However, in the 1980s, the word "exosomes" was proposed for small vesicles (30 - 100 nm) of endosomal origin. Theses exosomes were described by two independent groups studying reticulocyte maturation, they observed that small vesicles of endsomal origin were formed as multivesicular body (MVB) and released into the extracellular space as a consequence of fuse with the plasma membrane (PM)[5-7]. One decade later, the real biological significance of EVs in intercellular communication has been confirmed by mounting evidence. EVs were confirmed afterwards to be released by dendritic cells and B lymphocytes through a similar secretion pathway like exosomes, which could stimulate adaptive immune responses[8-9]. Advancing on these pioneering studies, EVs were not only released from many different cell types, such as epithelial cells[10], tumor cells[11], neurons[12], and mast cells[13], they were also found in diverse biological fluids, like breast milk[14], blood[15], urine[16], semen[17], and saliva[18]. Further description of EVs containing RNA, including microRNA, conformed that EVs sparked a strong interest as mediators of cell-to-cell communication[19-20]. It is currently believed that EVs serve as a means of extracellular communication via the exchange of lipids, proteins, and nucleic acids, play a key role in the regulation of physiological and pathological processes, such as tissue repair, blood coagulation, tumorigenesis, and immune surveillance[8, 21-25]. How EVs from different sources display multiple functions, however, is still unclear. Now it will be possible to answer this question, with widespread interest and enthusiasm in EV, reaching far beyond the EV research community, we are constantly obtaining detailed knowledge of the cellular biology of these vesicles from the increasing number of EV publications.

1.1 Definition

EVs are lipid bilayer-enclosed extracellular structures without functional nucleus and naturally released into the extracellular environment by an evolutionally conserved manner from almost all cells that range from different organisms such as prokaryotes to higher eukaryotes and plants[26-28].

1.2 Characteristics

EVs are nanometer sized vesicles encased in lipid membranes. The origin, nature, and features of these vesicles are diverse. EVs derived from different cells carry diverse cargoes including lipids and proteins, RNAs and DNAs. When in contact with recipient cells, the preassembled complex biological information of EVs that allow them to be targeted to recipient cells and elicits pleiotropic responses. Therefore, EVs can be used as important mediators of intercellular communication, affecting physiological and pathological conditions. And more notably, the accumulating data have indicated that EVs derived from different cells are different, possibly even from the same cells, showing that the membrane composition, content and size of EVs were highly heterogeneous and dynamic, which mainly depended on the cellular source, the state of cells, the environmental conditions of cell culture and other factors.

2 Biogenesis

With the in-depth study of the biogenesis of EVs, scientists can better understand the function of EVs and forecast the application of EVs in the future. It is generally acknowledged that distinct biogenesis pathways lead to different vesicle types. There are three main biogenesis of EVs, and as determined by biogenesis, EVs are derived into three classes: exosomes, microvesicles, and apoptotic bodies (Fig. 8 - 1).

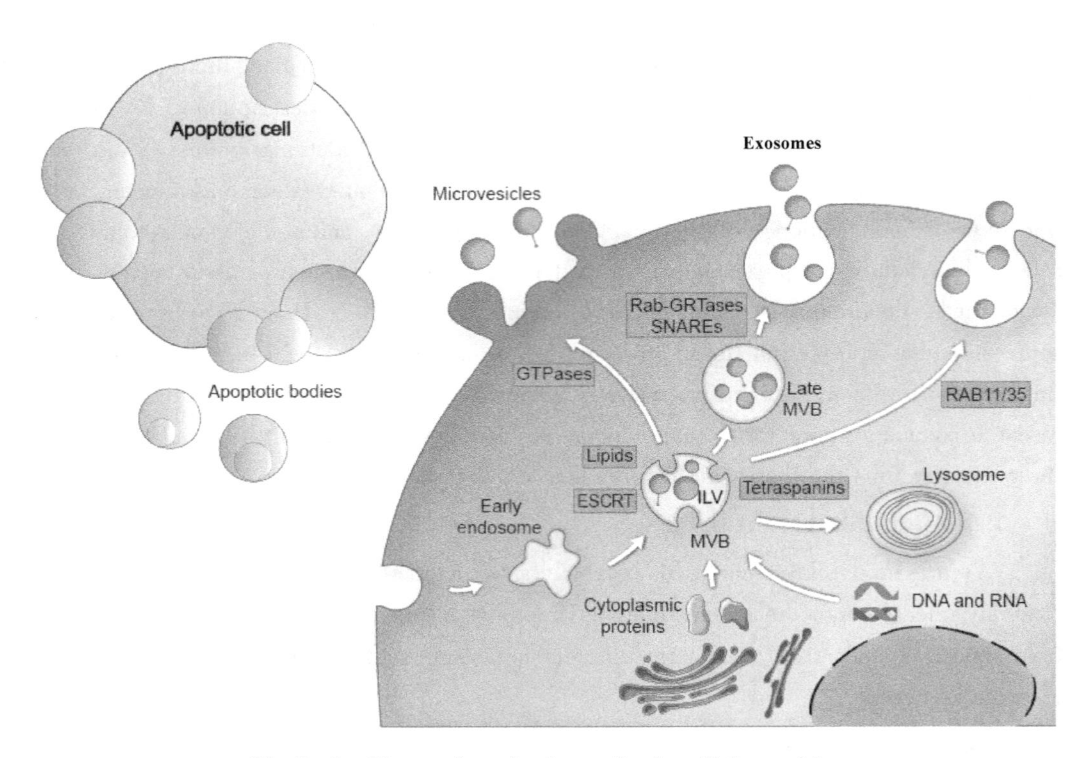

Fig. 8 - 1 Biogenesis and release of extracellular vesicles

Illustrated by 苏丹丹.

References:

[1] YÁNEZ-MÓ M, SILJANDER P R, ANDREU Z, et al. Biological properties of extracellular vesicles and their physiological functions [J]. Journal of extracell vesicles, 2015, 4: 27066.

[2] KOWAL J, TKACH M, THERY C. Biogenesis and secretion of exosomes [J]. Current opinion in cell biology, 2014, 29: 116 - 125.

Exosomes typically display a cup-like shape with a size ranging from 30 to 150 nm in diameter, and they are presumed to represent a homogeneous population[29]. Exosomes are derived from the intraluminal budding into early endosomes, and then formed as intraluminal vesicles (ILVs) by budding into intracellular MVBs. As the starting point of the biogenesis process, ILV formation, which is mediated by the endosomal sorting complexes required for transport (ESCRT) machinery and several molecules, such as tetraspanin microdomains and lipid (such as ceramide). In addition to being fused by lysosomes, the fate of MVBs can also be fused with plasma membrane (PM), which allows the release of ILVs as exosomes into the extracellular milieu[30]. Notably, distinct subpopulations of exosomes release from cells, with heterogeneous sizes and compositions, which elicit differential biological effects on the recipient cells.

Microvesicles (also called microparticles or ectosomes), in contrast to exosomes, are produced by outward budding and fission of the plasma membrane, which represent a more heterogeneous population with a size between 100 and 1 000 nm in diameter[31]. In response to stimuli, outward blebbing may be induced by multiple enzymes regulation and mitochondrial or calcium signaling, which selectively occur in the lipid-rich microdomains of the membrane, such as lipid rafts or caveolae domains, and then followed by cytoskeleton reorganization and phospholipid symmetry changes, microvesicles will eventually be released within a few seconds[24, 31]. Interestingly, evidence accumulated demonstrated that ESCRTs and tetraspanin microdomains also play a central role in microvesicles.

Other types of extracellular vesicles, apoptotic bodies, are exclusively released as blebs of cells undergoing apoptosis with caspase-mediated cleavage and activation of Rho-associated kinase I[32]. Apoptotic bodies showed significantly differences among cell types, ranging in size from 800 – 5 000 nm, larger than exosomes and microvesicles[31].

3 Classification

Based on cellular origin, biological features or particular biogenesis pathways, EVs are classified into various subtypes. Over the past decade, a number of publications have offered such a wide variety of definitions and names for many subtypes of EVs. Since there are still no specific markers available to distinguish subsets of EVs, and consensus has not yet emerged on the annotation of the subtypes of EVs, thus precise nomenclature for EVs is an ongoing problem that has plagued the EV-field. At present, according to minimal information for studies of extracellular vesicles 2018 (MISEV2018) launched by the International Society for Extracellular Vesicles (ISEV: www. isev. org/), EVs are classified and named from three main aspects.

(1) Physical characteristics: mainly according to the size range [<100 nm or < 200

nm (small vesicles, sEVs), >200 nm (medium and/or large vesicles, m/lEVs)] and density range (low, middle, high).

(2) Biochemical composition: for example, what molecules can be expressed in EVs, like $CD63^+/CD81^+$ – EVs, or stained EVs, like annexin A5-stained EVs, etc. .

(3) Conditions and origin: such as EVs from podocyte named as podocyte EVs, hypoxic EVs were term for EVs derived from hypoxic cells, and some EVs from the tumor cells were termed for large oncosomes, etc.[28].

However, it is still popular to divide EVs into three major subsets: ① apoptotic bodies, ② microvesicles, microparticles, or ectosomes, ③ exosomes[26, 33]. Until now, scientists have classified these EVs based on their content and size using a variety of preparation methods, such as differential centrifugation, microvesicles typically being isolated by centrifugation at 10 000 g to 20 000 g, and exosomes obtained by a very high speed centrifugation at or above 100 000 g. It is worth noting that these names, like exosomes and microvesicles, should be defined explicitly and prominently when used, as there have been various and contradictory definitions and inaccurate descriptions of unique biogenesis throughout history. Tab. 8 – 1 shows the main characteristics of different classes of EVs.

Tab. 8 – 1 Classification of EVs

Vesicle types	Characteristic				
	Origin	Size/nm	Morphology	Markers	Composition
Exosomes	MVB fusion with cell membrane	30 – 150	Cup-shaped	CD63, CD81, CD9, Tsg101, Alix	Protein, lipids, RNA, DNA
Microvesicles	Outward budding of cell membrane	100 – 1 000	Heterogeneous	CD40, integrins, selectins	Lipids, cell organelles, protein, RNA, DNA
Apoptotic bodies	Outward blebbing of apoptotic cell membrane	800 – 5 000	Heterogeneous	Caspase 3, histones	Cell organelles, nuclear fractions, proteins, RNA, DNA

4 Composition

In the past decades, since the discovery of EVs, numerous works have focused on exploring and providing exhaustive and comprehensive characterization of the contents of EVs, which specifically reflects the cellular origin, secretion mechanism, and localization of vesicles. It is worth noting that the important role of EVs in cellular communication depends on its unique biomolecule categories, such as proteins, lipids, nucleic acids, and sugars. EVs-mediated signals can be transmitted to multiple different messengers, even to remote cell, playing a vital physiological and pathological role.

4.1 Proteins

The proteins in different types of EVs obtained from cell cultures, tissue cultures or biofluids are commonly analyzed by proteomic analysis, protein staining, immunoblotting, antibody-coupled bead flow cytometry analysis and immuno-gold labelling combined with electron microscopy, which revealed the specific subset of proteins from endosomes, the cell membrane and the cytosol, but mostly absent from intracellular organelles (nucleus, endoplasmic reticulum, mitochondria, Golgi). The observation of protein from EVs have been made available and accessible through the creation of public on-line databases: EVpedia (www.evpedia.info)[34], Vesiclepedia (www.microvesicles.org/)[35], and ExoCarta (www.exocarta.org)[36]. Of notes, highly purified EVs used to study protein components should be devoid of pollutants, such as proteins from serum in the culture medium and protein components from intracellular compartments (such as the endoplasmic reticulum or mitochondria) that have never been in contact with EVs. Interestingly, the protein composition and protein post-translational modifications of EVs are, in some instances, associated with the cell types, vesicle trafficking, and mode of biogenesis. Some types of EVs, such as exosomes originating from the endosomal compartment, they contain protein, like members of the ESCRT complexes (Alix and Tsg 101), that are involved in the biogenesis of MVB; and they are rich in endosome-associated proteins (annexins, Rab GTPase, SNAREs and flotillin) that are important for transport and fusion[37-38]. Membrane proteins that are known to not only cluster at the plasma membrane and endosomes, and are also enriched on the membrane of EVs. These membrane proteins include MHC molecules, which are involved in antigen presentation; and tetraspanins, a family of > 30 proteins consisting of four transmembrane domains, such as CD37, CD53, CD63, CD81, and CD82, were previously considered to be specific markers for exosomes, but these proteins have also been detected in apoptotic bodies and microvesicles[26]. Cytosolic proteins, specific stress proteins (heat shock proteins) and signaling proteins are also found in EVs[26]. In addition, owing to plasma membrane origin, certain types of EVs, such as microvesicles, tend to be enriched in different proteins, including integrins, P-selectin and glycoprotein Ib (GPIb), as well as carrying more proteins with posttranslational modifications, such as glycoproteins or phosphoproteins[26]. While the protein profiles of different EV subgroups show a substantial overlap, nevertheless, certain proteins are more enriched in one than in other EV subtypes. And it is unclear whether this overlap is at least partially due to the applied separation techniques, which have not yet been able to achieve complete separation of the EV subtype, and protein aggregates. There is, in addition, one further point to note that currently no specific markers that uniquely identify individual EV subtypes.

4.2 Lipids

Interestingly, although the studies of the lipid content of EVs were not as frequent as the protein analyses, the investigation conformed that the lipids of EVs act as bridges between cells. When comparing cells with EVs derived from them, the membranes of EVs are consist of a lipid bilayer which similar to cell plasma membrane. Interestingly, the lipid-bilayered EVs can protect their genetic cargo from being digested by enzymes. However, all studies showed the particular organization and lipid composition of EVs are distinct from the parent cell, even though they share common features. The lipid content of EVs are generally enriched in phosphatidylserine, sphingomyelin, cholesterol, ceramide, and glycosphingolipids, which confer an EVs structure that is close to detergent-resistant subdomains of the plasma membrane called lipid rafts. In contrast to cellular membranes, some types of EVs, such as exosome, contain more phosphatidylserine in the outer surface, which represents one of the characteristic EVs markers and may facilitate their internalization by recipient cells. Interestingly, some studies have reported phosphatidylserine deficiency in microvesicles, which may indicate that phosphatidylserine externalization is not the only mechanism for EVs biogenesis[39]. However, the lack of phosphatidylserine positive staining may be due to the inability of current methods to detect low expression of phosphatidylserine, or to the fact that phosphatidylserine bind to protein S, Del-1, or lactadherin, thereby making it unrecognizable[31, 39]. Some studies have also revealed an increased fraction of sphingolipids, cholesterol and disaturated lipids in EVs compared with that in cell membranes, which implies that the lipid bilayer of EVs exhibit greater rigidity than cell membrane and may contribute to increase resistance to degradation and stability as a carrier for various biomolecules. Several lipids in EVs have been described as key mediators of signaling in activated receptor cells, such as fatty acids, cholesterol and eicosanoids. In addition, compared with cell membrane, membrane of EVs exists more frequent transmembrane flip-flop lipid movements that can promote the exchange of the inner and outer leaflet of the membrane, so lipids are emerging as important factors that constitutes the biological function of EVs[40].

4.3 Nucleic Acids

In addition to proteins and lipids, a major breakthrough in the field was the demonstration that EVs carry abundant genetic material, such as messenger RNA (mRNA) and microRNAs (miRNA), and that EVs-related mRNAs can be translated into proteins by target cells[19, 41]. EVs are treated by RNaseA to verify whether RNAs are within the cytosolic lumen or associated with the outer membrane of EVs. The RNA enclosed in EVs may exist in the form of freely circulation or be bound to the protein complexes, which is predominantly smaller than 700 nucleotides (nt) in size, while the average size of cellular mRNA varies between 400 and 12 000 nt[42]. Recently, total RNA from EVs have been demonstrated by unbiased deep sequencing approaches that, in addition to mRNA, mRNA fragments and miRNA. EVs also contain a large variety of other RNA species, including ribosomal RNA (rRNA), long non-coding RNA, RNA transcripts overlapping with protein coding regions, piwi-interacting RNA, repeat sequences, structural RNAs, fragments of tRNA-, vault- and Y-RNA, and small interfering RNAs. Although various RNA types have been identified by sequencing in EVs, their selective packaging and function in recipient cells have yet to be determined. Interestingly, although some profiles of EV-RNA do not consistent with those of cellular RNA, EVs and cells share many transcripts, and some RNAs isolated from EVs were found to be systematically enriched in the originating cells, which may indicate that RNA molecules in cells have been selectively incorporated into EVs. The types and amounts of RNA in different EVs are generally various, and this variability may depend on the cell type of origin. ln addition, different RNA isolation methods lead to significant differences in the yields and patterns of EV-RNA, which largely depends on the experimental variations between studies and the lack of quantitative data. And it is important to note that extracellular RNA exists in different forms, but many studies have failed to show whether the identified extracellular RNAs were truly associated with EVs or RNA-protein complexes. It is noteworthy that after contacting with the receptor cells, EV will release specific RNA molecules, which may be translated into proteins or compete with cellular RNA for miRNAs or RNA-binding proteins, thus regulating stability and translation, with intrinsic influence on the regulation of gene expression in the recipient cells.

In contrast to the in-depth study of RNA, the presence of DNA in EVs has so far been little explored and much remains unclear. However, there is compelling evidence from some studies reporting DNA types in EVs that contain genomic DNA, such as single-stranded DNA and double-stranded DNA (dsDNA), in addition to mitochondrial DNA (mtDNA) and oncogene amplification (c-Myc). It is clear that different EV subsets carry different DNA cargoes that perform different biological functions. Some studies have shown that DNA carried by tumor-derived EVs, such as dsDNA, can reflects the mutational status of parent tumor cells and the genetic status of tumour, such as the amplification of oncogene c-Myc, which demonstrate its great potential as a biomarker. In addition, studies have shown that the migration of DNA may take place via EVs, for example, mtDNA can enter into other cells through EVs, which is conducive to the spread of various pathologies.

In conclusion, EVs containing many types of nucleic acids, due to their unique biological structure, may serve as nucleic acid delivery vectors, and represent a new mechanism for the exchange of genetic material between cells.

5 Biological Functions

The significant effect of EVs exert on physiological and pathological processes in a pleiotropic manner that lies in their capacity to interact with recipient cells and transfer information through their extensive and variable contents of proteins, lipids, and nucleic acids. EVs can be taken up by recipient cells via different mechanisms (Fig. 8 – 2). EVs may direct interact with recipient cells via: ① direct membrane fusion with recipient cell plasma membrane; ② lipid raft-, clathrin-, and calveolae-dependent endocytosis; ③ micropinocytosis or phagocytosis. In these ways, effectors including transcription factors, functional protein, infectious particles, oncogenes and non-coding regulatory RNAs are introduced into receptor cells to participate in the maintenance of normal physiology and the regulation of disease

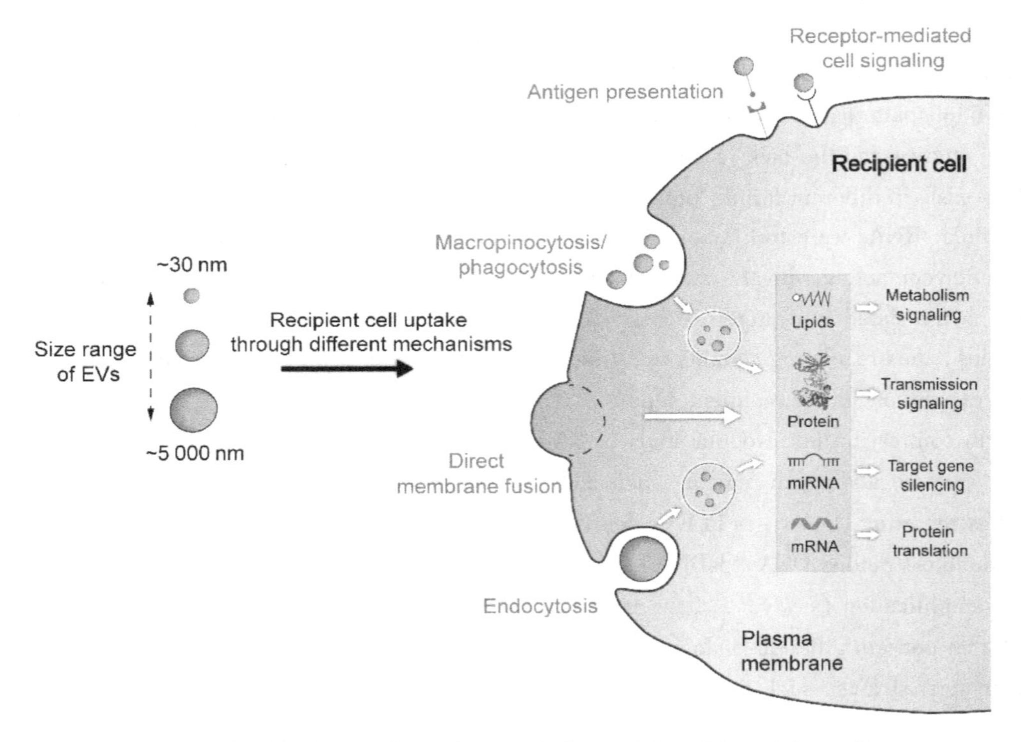

Fig. 8 – 2 The interactions of extracellular vesicles with recipient cells

Illustrated by 苏丹丹.

References:

[1] EL ANDALOUSSI S, MAGER I, BREAKEFIELD X O, et al. Extracellular vesicles: biology and emerging therapeutic opportunities [J]. Nature reviews drug discovery, 2013, 12: 347 – 357.

[2] ZABOROWSKI M P, BALAJ L, BREAKEFIELD X O, et al. Extracellular vesicles: composition, biological relevance, and methods of study [J]. Bioscience, 2015, 65: 783 – 797.

pathogenesis.

5.1 Immunoregulation

In addition, EVs can also communicate with recipient cells by ④ receptor-mediated cell signaling and ⑤ antigen presentation through ligand/receptor molecules interaction on their respective surfaces, such as transfer of both MHC molecules and antigens, which help to target specific cell types and play an important role in immune regulation. Based on the unique biological structure and complex biological functions of EVs, EVs can thus be regarded as promising signalosomes: multifunctional signaling complexes that control basic cellular behavior and biological functions.

Various functions of EVs have been widely reported, some of which are well documented to be related to some form of immune regulation (Fig. 8-3). EVs from both immune (e.g. dendritic cells, macrophages, and B cells) and nonimmune cells (e.g. mesenchymal stem cells and endothelial cells, tumor cells) are contributed significantly to antigen-specific and nonspecific immune regulation through several distinct mechanisms. EVs carry a similar pattern of surface immune regulatory proteins that modulate immune activity by direct contact with T cells and other immune cells. In addition, antigen-presenting cells (APCs) such as DCs and B cells increase the release of EVs after cognate T cells interactions, suggesting that EVs are essential for conferring partial immunomodulatory effects. Interestingly, EVs can regulate immune effect by indirect presentation, they can be internalized by the APCs, where may allow the peptide/MHC complexes of EVs to be presented to T cells. In addition, EVs also can be degraded within the APCs, and the peptides of EVs used as a source to loaded onto the vesicle-MHC complex and further processed for presentation on the surface of the APCs. Furthermore, EVs uptaken by APCs can be subsequently secreted as APC-derived EVs carrying p-MHC Ⅱ, which are able to regulate immune responses. Without internalization, EVs can retain on the APC surface, where they are decorated with costimulatory molecules and other regulatory molecules, and then directly present p-MHC complexes to interact with T cells (cross-dressing). Alternatively, EVs interact with or are absorbed by APCs, leading to increased production of cytokines, such as IL-10 and TGF-β1, thereby enabling them to activate antigen-specific immune responses. Therefore, EVs have the potential to mediate immune responses in a tolerogenic manner. For example, EVs derived from inmmune cells, such as B cell and DCs, have the ability to carry MHC Ⅱ, antigens, costimulatory molecules, and other proteins, and can promote the activation of $CD4^+$ T cells and $CD8^+$ T cells, thereby triggering the immune response. By contrast, EVs can also confer immune suppression via several mechanisms: Treg-derived EVs containing CD73 can enhance immunosuppression by suppressing natural killer (NK) and $CD8^+$ cell. T cell-derived EVs can inhibit the differentiation of monocytes into DCs and prevent DCs maturation. Furthermore, T cell-derived EVs can also suppress the immune response through negative regulation by carrying co-stimulators such as PD-L1 and CTLA-4. In summary, the multiple abilities of different types of EVs play an important role in mediating immune-related diseases, such as cancer, systemic/chronic inflammatory, and autoimmune diseases.

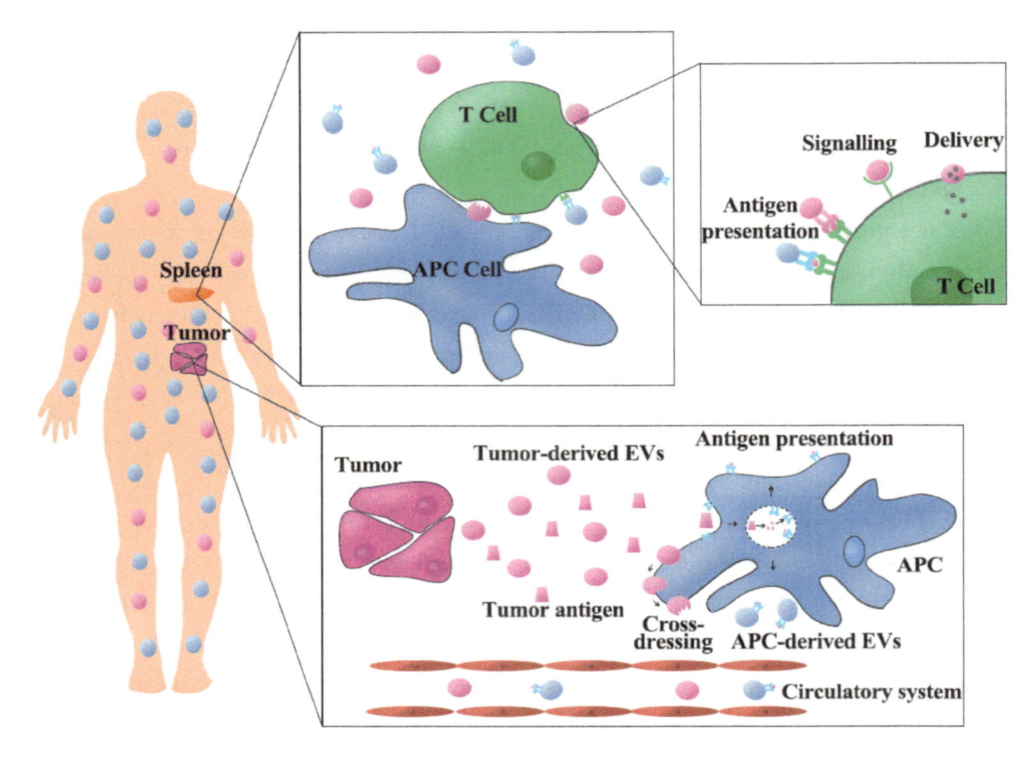

Fig. 8 – 3 Role of endogenous EVs in regulating immune responses

Illustrated by 苏丹丹.

References:

ROBBINS P D, DORRONSORO A, BOOKER C N. Regulation of chronic inflammatory and immune processes by extracellular vesicles [J]. Journal of clinical investigation, 2016, 126 (4): 1173 – 1180.

5.2 Tumorigenesis

Numerous studies have implicated that EVs participate in the dissemination of that tumor- and stroma-derived EVs play an important role in different physical processes of the metastatic cascade. EVs have the ability to induce the proliferation of cancer cell, which will eventually stimulate the growth of the tumor. In particular, EVs can drive tumor cell metastasis. By deposing ECM components such as fibronectin into EVs and bind to integrins present on EVs, EVs provide a substrate favoring cell adhesion, promote effective and directionally persistent migration, and enhance directional cancer cell motility. Moreover, the EVs containing metalloproteinases such as MT1-MMP not only participate in the formation of invadopodia and promote cell motility, but also contribute to remodeling of ECM. Some evidence illustrate that cancer-derived EVs can alter the cellular physiology of both surrounding and distant non-tumor cells by triggering vascular permeability, and then it will create a fibrotic environment and promote the migration of bone marrow progenitor cells to pre-metastatic niche, and ultimately trigger tumor growth. In the tumor microenvironment, EVs carry costimulatory molecules that inhibit the immune response, such as PD-L1 and CTLA4, which can enhance immune escape and thus protect cancer cells by regulate the killing activity of T cell. Collectively, EVs have an important and fundamental role in various steps relate to tumor progression and tumor-associated pathologies such as thrombotic events.

5.3 Angiogenesis

Angiogenesis is strictly regulated by a precise balance between stimulation and inhibition signals to maintain a stable internal environment in organism. An increasing number of studies show that EVs derived from different types of cells, such as ECs, endothelial colony-forming cells, MSCs, erythrocytes, leucocytes, and platelets, can actively act as vectorial signalosomes to transmit and influence the stimulation and inhibition signals in the angiogenic programs, and ultimately affect vessel formation on the key steps. As dynamic systems, in response to changes in the microenvironment, EVs may exert their beneficial or detrimental effects on angiogenesis, depending on cellular source, production conditions, and content. Notably, under specific conditions, such as hypoxia and angiogenic growth factor stimulation, EVs with angiogenic potential are favored to be released, which can deliver powerful proangiogenic signals by transferring microRNAs (miR-31, miR-125a, miR-126, miR-150, miR-214, and miR-296), proteins (VEGF, FGF-2, PDGF, c-kit, SCF, RANTES, CD40L, CRP, metalloproteases, uPA, and Sonic hedgehog), lipids (S1P), and transcription factors (STAT3, STAT5) to receptor cells or by activating key signaling pathways (PI3K, ERK1/2, Wnt4/β-catenin, and NF-κB) to facilitate angiogenesis[31]. However, some recent studies have shown that EVs from ECs, platelets, or lymphocytes can also exert an inhibitory activity on vascular growth, triggered primarily by increasing the oxidative stress in target cells and CD36-dependent uptake of EVs[31]. Thus, EVs also mediate abnormal vascular growth, which will become the cause of triggering many diseases, such as cancer, atherosclerosis, ischemic disease, or rheumatoid arthritis. Taken together, EVs are widely documented to induce either pro- or anti-angiogenic signaling and are involved in vascular development, growth, and maturation, and their potential therapeutic applications in regenerative medicine or angiogenation-related diseases are attracting increasing attention.

5.4 Neuroregulation

In the brain, in addition to classical synaptic neurotransmission, neurons communication through the basal release and uptake of EVs from surrounding cells. These EVs are also released into the cerebrospinal fluid (CSF) and blood, allowing longer-range communication within the central nervous system (CNS) and contributing to a range of neurobiological functions (including synaptic plasticity). Given their small size and bilayer structure, EVs can move from the site of release by diffusion and cross the blood-brain barrier (BBB), and eventually reach several biological fluids. It is worth noting that EVs can be secreted from microglia, oligodendrocytes, astrocytes, and MSCs, which means that they are able to carry different cargos, such as nucleic acids, cytokines, transcription factors, cytosolic proteins, Ab scavenger enzymes, Ab and tau protein, therefore they have the potential of positive and negative influence on neurons. Depending on the condition of the donor cell type, EVs are emerging as clinically valuable markers serve to monitor the outcome of neurodegenerative conditions. Some evidence shows that cells use EVs to remove toxic proteins and aggregates from their cytoplasm. Moreover, it also elegantly demonstrated that EVs serve as potential carriers to interact with healthy cells and delivery misfolded proteins that associate with various neurological disorders. For example, multiple studies have shown that EVs are involved in the pathogenesis of AD, EVs are involved in the transfer of pathogenic amyloid-β peptides deposition in the parts of the brain. Similarly, EVs containing α-synuclein protein facilitate the local spread of Parkinson's disease from enteric neurons to the brainstem and higher cortical centers. In addition, EVs can also carry superoxide dismutase (SOD) 1 to aggravate the amyotrophic lateral sclerosis, as well as load with huntingtin to enhance Huntington's disease. However, EVs released by MSCs also have a protective effect on AD: they can reduce both intracellular and extracellular Aβ levels after their internalization in N2a cells overproducing Aβ.

6 Application

6.1 Drug Delivery

Indeed, it is clear that EVs serve important roles in cellular communication by delivering message to the target cells in a paracrine or endocrine manner. Because EVs are biocompatible, immunologically inert, and its natural structure confer them the innate ability to cross biological barriers including the BBB, so the most intriguing potential feature of EVs are being explored as natural vectors to encapsulate and convey different cargoes, including various bioactive small molecules and drugs with suboptimal pharmaceutical properties, which can further transmit them to neighboring or more distant cells, mediating various signaling cascades. The fact that EVs are secreted by most cells, are rich in proteins and different RNA species, which are able to transfer their content to recipient cells. Although a lot of studies have definitely demonstrated that EVs may be highly suitable candidates for drug delivery, particularly therapeutic nucleic acid delivery, it must be noted that there are problems with low cellular uptake, off-target toxicity, suboptimal pharmacokinetics, or poor stability. In order to enhance the ability of EVs as drug delivery carriers and compensate for their deficiencies, various techniques have been explored to load therapeutic cargo into EVs by manipulating EVs themselves or parent cells. These techniques can be divided into two basic approaches: exogenous loading and endogenous loading. Exogenous modification occurs after EVs collection to incorporate the desired therapeutic cargo (small molecules/proteins/RNA) into or onto EVs through various manipulations including co-incubation, electroporation, permeabilization using saponin, freeze-thaw cycles, extrusion, and sonication. For example, many studies have shown that EVs carrying small molecule, such as curcumin, doxorubicin, paclitaxel, and other chemotherapeutic drug can improve pharmacokinetic profiles, effective cargo delivery, as well as retention in tumor cells, compared to liposomes and polymer-based synthetic nanoparticles[43]. In contrast to exogenous loading, endogenous loading implies that cargo is expressed in the producer cell through incubating and genetically modifying to overexpress desired RNAs, proteins, or small molecules of interest (with or without modification to promote packaging) in parent cells, and then desired cargo will be naturally incorporated into the secreted EVs during their biogenesis. There are many emerging engineering strategies for modifying parental cells to produce EVs with the targeted RNAs and proteins that can be generated potential therapeutic candidates for antitumor, neurogenesis, immune regulation, and tissue repair. Before the full potential of EVs to delivery drug can be realized, the cell types which EVs are derived need to be evaluated, a broader range of potential drug cargoes explored, drug-loading procedures optimized and standardized and tissue-targeting and barrier-crossing optimized to further enhance the potency of delivery to the desired tissues.

6.2 Therapeutic Tools

By virtue of the biogenesis and bioactive contents of EVs and their actions in cellular communication, EVs may be a key mechanism and vehicle with tremendous potential as therapeutic agents. They are likely to play a role in altering cellular signaling, modulating immune response, reducing inflammation, and leading to tissue repair. Thereby they have the ability to apply in diverse areas, such as cell-free cancer immunotherapy, modulating inflammatory and autoimmune diseases, and treating infections caused by viruses and microbial pathogens, and regenerative medicine.

In particular, EVs derived predominantly from stem cells have the potential to drive tissue regeneration by recruiting and/or reprogramming required cells to the injured sites, suppressing healthy cell apoptosis and stimulating cell proliferation, inducing angiogenic programmes in quiescent endothelial cells, and delivering immunomodulatory signals.

EVs can be acquired from the bone marrow or peripheral blood, such as MSCs and HSCs. A growing number of animal models and ongoing clinical trials have confirmed that MSC-based therapies are successful to repair injured tissue, provided that stem cells can be transplanted into injured tissues and subsequently differentiate to replace injured cells. Many studies have shown that MSCs rely on the release of EVs and their cargoes to mediate paracrine effect and protect tissue from ischaemia. They have been applied to treat various diseases, not only cardiovascular disease but also graft versus host disease and Crohn's disease, as well as acute kidney injury.

There has been a rapid increase in the number of studies indicating that EVs are derived from specific differentiated cell types could participate in the activation of the immune system, suppression of inflammation, and augmentation of antigen presentation in a tolerogenic manner. This may be due to the fact that EVs can protect T cells from activation-induced cell death, induce macrophages to release pro-inflammatory cytokines and tumor necrosis factors, and enhance NK cell activity. Conversely, in light of the source and status of the EVs, it is not surprising that EVs have properties of immunosuppressive that inhibit T cell activation, differentiation and proliferation, favour the expansion of Treg cell subsets and myeloid suppressor cells, induce FASL-mediated T cell apoptosis, and inhibit inflammation via inflammatory factor delivery. Therefore, the innate ability of EVs to modulate immune responses could be exploited for immunotherapy, particularly for inflammatory and autoimmune diseases, such as arthritis, diabetes, and lupus, but also in the treatment of infectious diseases and cancer.

Due to the fact that EVs are considered as vectorial signalosomes and bioactive cargoes, they also have innate therapeutic potential in other diverse diseases, such as cancer, infectious diseases and neurodegenerative disease. The naturally derived EVs are also be exploited for the delivery of exogenous therapeutic reagents, such as anti-inflammatory drugs and vaccine-like tumor-associated antigens. Apart from being used as therapeutics in recent clinical and preclinical investigations, EVs containing RNAs, lipids, and other proteins may also serve as diagnostic, prognostic, and predictive biomarkers.

6.3 Inhibiting Extracellular Vesicles in Disease

In fact, the growing evidences show that EVs are involved in many pathological conditions. For example, the level of circulating EVs is strongly related with many aspects of tumorigenesis and tumour-related pathology. Therefore, reducing the load of circulating EVs or specific inhibition of EVs provides a potential strategy for reducing disease progression. There are some certain strategies for inhibiting EVs that involve inhibiting various aspects of EVs function, inhibiting biogenesis or releasing of EVs, or targeting specific EVs components and inhibiting their uptake (Fig. 8 – 4).

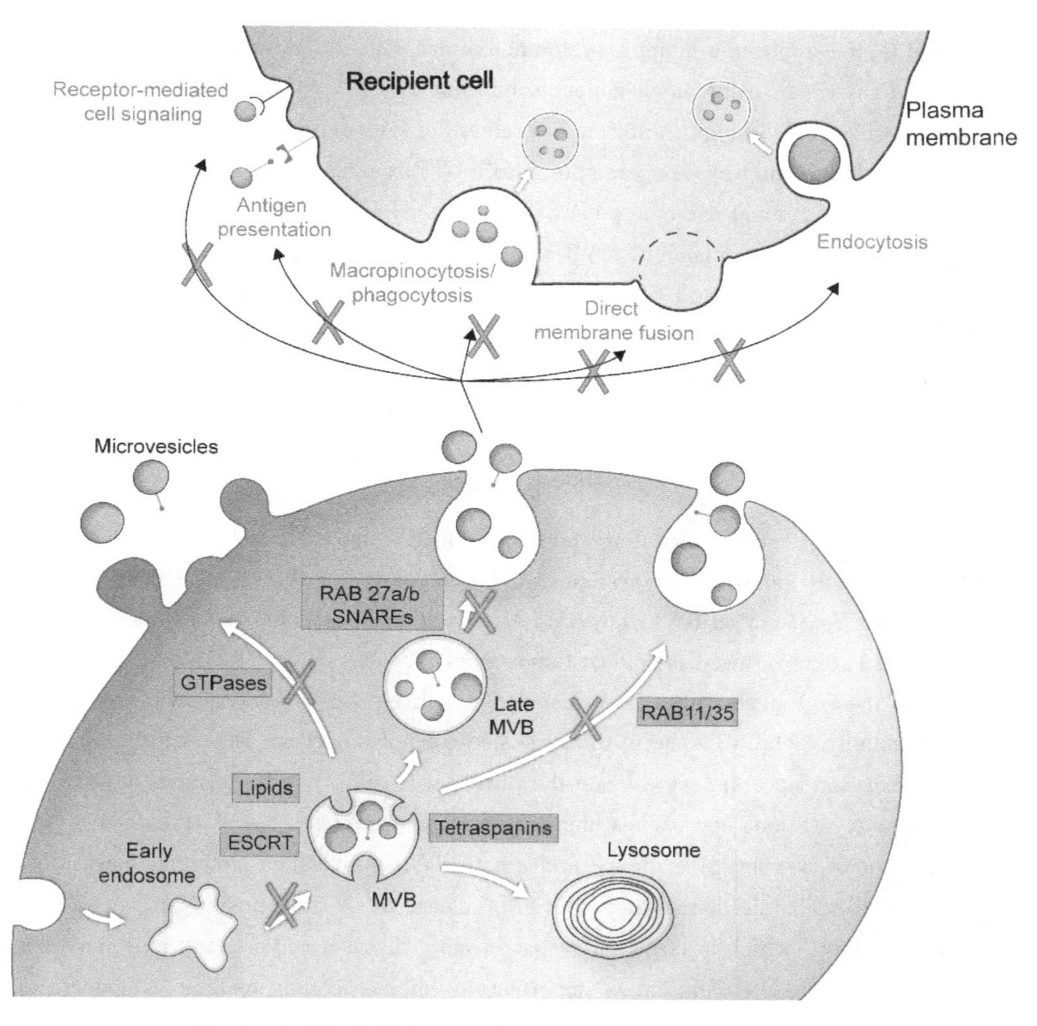

Fig. 8 – 4 The inhibitions of EVs uptaken by recipient cells

Illustrated by 苏丹丹.

References:

[1] EL ANDALOUSSI S, MAGER I, BREAKEFIELD X O, et al. Extracellular vesicles: biology and emerging therapeutic opportunities [J]. Nature reviews drug discovery, 2013, 12: 347 – 357.

[2] SUN Z, WANG L, DONG L, WANG X. Emerging role of exosome signalling in maintaining cancer stem cell dynamic equilibrium [J]. Journal of cellular and molecular medicine, 2018, 22 (8): 3719 – 3728.

The production of EVs can be lowered by blocking the components that are crucial for the formation of EVs, such as tetraspanins and ceramide, which are involved in exosome biogenesis. For example, small-molecule inhibitors of neutral sphingomyelinase can be used to block ceramide formation in cells. Similarly, blood-pressure-lowering drug amiloride (which generally attenuates endocytic vesicle recycling) also can block the secretion of EVs. In addition, the formation of EVs can also be attenuated by direct interference of syndecan proteoglycans and their cytoplasmic adaptor syntenin interact with the exosomal ALIX using RNA interference (RNAi) or using small-molecule inhibitors.

Some studies indicate that inhibiting the release of EVs depends on components that involve in the release pathway. For example, Rab GTPases family, including RAB27A/B, RAB11 and RAB35, might serve as potential targets for inhibiting the release of EVs by impairing the docking and/or fusion of MVB with the cell membrane.

Alternatively, since several EV uptake mechanisms have been proposed, it may be advantageous to interrupt EV uptake in recipient cells by blocking EVs ligands or cell surface receptors involved in EV binding or internalization. For example, phosphatidylserine exposure on cellular membranes and the membrane of EVs, which is important for cell adhesion, so in order to reduce the uptake of EVs released from tumor cells, diannexin can be applied to block phosphatidylserine[44].

In addition, the specific signaling components of EVs have been shown by some studies to be therapeutically relevant. For example, it has been shown that several drugs targeting components of EVs, such as RNAi targeting MET and FASL-specific monoclonal antibodies targeting FASL1, can be used to reduce tumor growth[45-46].

Some of these strategies have been tested, providing proof-of-concept evidence for their potential feasibility. Notably, some of the presented examples may not be specific towards EV-mediated effects but nevertheless indicate the principal feasibility of these types of strategies.

In summary, various approaches highlighted above have been tested, they are attractive by providing proof-of-concept evidence and are likely to prove their potential feasibility by developing small-molecule therapeutics. But to be clear, these strategies are still in early stages of development, which is largely untested in more disease models, and may have many unknown side effects. Therefore, it is important to emphasize that specific components involved in EVs biogenesis, such as proteins, are also important in several other core cellular processes to regulate the normal biological processes, and therefore interference with EV biogenesis may lead to undesirable off-target effects. In addition, it is to be noted that such approach may require a specific drug delivery system that can specifically target the populations of cells that produce specific EVs.

7 Looking Forward

Over the last few decades, intense research within the field of EVs has increased our understanding of the biogenesis, molecular content, and biological function of EVs. In light of superior biological structure, EVs have the potential to overcome biological barriers to free transport in body fluids and to recipient cells in remote tissues. As important conveyers, EVs can deliver information between cells via the transmission of various proteins, bioactive lipids and genetic information, and they may eventually alter the phenotype and function of recipient cells. Thus, multiple observations have described the impact of EVs in numerous biological and pathological processes. And EVs are emerging as highly potent therapeutic entities in diverse diseases. However, there are still hurdles to overcome before using EVs as therapies. Depending on the intended therapeutic use, it is critical to choose the appropriate source of EVs, which will affect the outcome of the final treatment. Another aspect that needs to be addressed is the exploration of the optimal method that allows for isolating large-scale purified and clinical-grade EVs. In addition, the storage and stability of EVs as an off-the-shelf therapy must be further examined. Critically, further characterization of therapeutic EVs in relevant preclinical models is needed to assess their safety, toxicology, and pharmacokinetic and pharmacodynamic profiles, which will help support clinical application prediction and dose determination. Given the enormous role, EV plays in pathological conditions, there are many strategies that can be applied to directly target EVs and inhibit their deleterious effects mediated in disease. Their inherent therapeutic potential can not only stimulate regenerative responses, but also serve as drug delivery vectors to deliver nucleic acids, therapeutic proteins and small-molecule drugs that combine with targeting moieties. The field is still in its infancy, but research on how to optimize large-scale production of EVs and dissect their complex biogenesis, content, and biological function is still ongoing. It is likely that EVs will represent a platform between drug delivery, biologics, and cell therapies that will be fully utilized in highly potent multifaceted biopharmaceuticals in the future.

Supplement

List of Abbreviations	
AD	Alzheimer's disease
APCs	antigen-presenting cells
BBB	blood-brain barrier
CNS	central nervous system
CSF	cerebrospinal fluid
DCs	dendritic cells
dsDNA	double-stranded DNA
ECM	extracellular matrix
ECs	endothelial cells
ESCRT	endosomal sorting complex required for transport

(To be continued)

List of Abbreviations	
EVs	extracellular vesicles
ILVs	intraluminal vesicles
MHC Ⅱ	major histocompatibility complex Ⅱ
miRNA	microRNAs
mRNA	messenger RNA
MSCs	mesenchymal stem cells
mtDNA	mitochondrial DNA
MVB	multivesicular body
nt	nucleotides
PD	parkinson's disease
PM	plasma membrane
rRNA	ribosomal RNA
SOD	superoxide dismutase
tRNAs	transfer RNA

References

[1] CHARGAFF E, WEST R. The biological significance of the thromboplastic protein of blood [J]. Journal of biological chemistry, 1946, 166: 189 - 197.

[2] WOLF P. The nature and significance of platelet products in human plasma [J]. British journal of haematology, 1967, 13: 269 - 288.

[3] ANDERSON H C. Vesicles associated with calcification in the matrix of epiphyseal cartilage [J]. Journal of cell biology, 1969, 41: 59 - 72.

[4] TRAMS E G, LAUTER C J, SALEM N J R, et al. Exfoliation of membrane ecto-enzymes in the form of micro-vesicles [J]. Biochimica et biophysica acta, 1981, 645: 63 - 70.

[5] HARDING C, HEUSER J, STAHL P. Receptor-mediated endocytosis of transferrin and recycling of the transferrin receptor in rat reticulocytes [J]. Journal of cell biology, 1983, 97: 329 - 339.

[6] PAN B T, TENG K, WU C, et al. Electron microscopic evidence for externalization of the transferrin receptor in vesicular form in sheep reticulocytes [J]. Journal of cell biology, 1985, 101: 942 - 948.

[7] JOHNSTONE R M, ADAM M, HAMMOND J R, et al. Vesicle formation during reticulocyte maturation. Association of plasma membrane activities with released vesicles (exosomes) [J]. Journal of biological chemistry, 1987, 262: 9412 - 9420.

[8] RAPOSO G, NIJMAN H W, STOORVOGEL W, et al. B lymphocytes secrete antigen-presenting vesicles [J]. Journal of experimental medicine, 1996, 183: 1161 - 1172.

[9] ZITVOGEL L, REGNAULT A, LOZIER A, et al. Eradication of established murine tumors using a novel cell-free vaccine: dendritic cell-derived exosomes [J]. Nature medicine, 1998, 4: 594 - 600.

[10] VAN NIEL G, RAPOSO G, CANDALH C, et al. Intestinal epithelial cells secrete exosome-like vesicles [J]. Gastroenterology, 2001, 121: 337 - 349.

[11] WOLFERS J, LOZIER A, RAPOSO G, et al. Tumor-derived exosomes are a source of shared tumor rejection antigens for CTL cross-priming [J]. Nature medicine, 2001, 7: 297 - 303.

[12] FAURÉ J, LACHENAL G, COURT M, et al. Exosomes are released by cultured cortical neurones [J]. Molecular and cellular neuroscience, 2006, 31: 642 - 648.

[13] SKOKOS D, GOUBRAN-BOTROS H, ROA M, et al. Immunoregulatory properties of mast cell-derived exosomes [J]. Molecular immunology, 2002, 38: 1359 - 1362.

[14] ADMYRE C, JOHANSSON S M, QAZI K R, et al. Exosomes with immune modulatory features are present in human breast milk [J]. Journal of immunology, 2007, 179: 1969 - 1978.

[15] CABY M P, LANKAR D, VINCENDEAU-SCHERRER C, et al. Exosomal-like vesicles are present in human blood plasma [J]. International immunology, 2005, 17: 879 - 887.

[16] PISITKUN T, SHEN R F, KNEPPER M A. Identification and proteomic profiling of exosomes in human urine [J]. Proceedings of the National Academy of Sciences of the United States of America, 2004, 101: 13368 - 13373.

[17] POLIAKOV A, SPILMAN M, DOKLAND T, et al. Structural heterogeneity and protein composition of exosome-like vesicles (prostasomes) in human semen [J]. Prostate, 2009, 69: 159 - 167.

[18] OGAWA Y, MIURA Y, HARAZONO A, et al. Proteomic analysis of two types of exosomes in human whole saliva [J]. Biological & pharmaceutical bulletin, 2011, 34: 13 - 23.

[19] VALADI H, EKSTRÖM K, BOSSIOS A, et al. Exosome-mediated transfer of mRNAs and microRNAs is a novel mechanism of genetic exchange between cells [J]. Nature cell biology, 2007, 9: 654 - 659.

[20] RATAJCZAK J, MIEKUS K, KUCIA M, et al. Embryonic stem cell-derived microvesicles reprogram hematopoietic progenitors: evidence for horizontal transfer of

mRNA and protein delivery [J]. Leukemia, 2006, 20: 847 –856.

[21] LEE Y, EL ANDALOUSSI S, WOOD M J. Exosomes and microvesicles: extracellular vesicles for genetic information transfer and gene therapy [J]. Human molecular genetics, 2012, 21: R125 – R134.

[22] BOBRIE A, COLOMBO M, RAPOSO G, et al. Exosome secretion: molecular mechanisms and roles in immune responses [J]. Traffic, 2011, 12: 1659 – 1668.

[23] RAK J, GUHA A. Extracellular vesicles—vehicles that spread cancer genes [J]. Bioessays, 2012, 34: 489 –497.

[24] DEL CONDE I, SHRIMPTON C N, THIAGARAJAN P, et al. Tissue-factor-bearing microvesicles arise from lipid rafts and fuse with activated platelets to initiate coagulation [J]. Blood, 2005, 106: 1604 – 1611.

[25] GATTI S, BRUNO S, DEREGIBUS M C, et al. Microvesicles derived from human adult mesenchymal stem cells protect against ischaemia-reperfusion-induced acute and chronic kidney injury [J]. Nephrology dialysis transplantation, 2011, 26: 1474 – 1483.

[26] YÁÑEZ-MÓ M, SILJANDER P R, ANDREU Z, et al. Biological properties of extracellular vesicles and their physiological functions [J]. Journal of extracellular vesicles, 2015, 4: 27066.

[27] GYÖRGY B, SZABÓ TG, PÁSZTÓI M, et al. Membrane vesicles, current state-of-the-art: emerging role of extracellular vesicles [J]. Cell mol life sci, 2011, 68: 2667 –2688.

[28] THÉRY C, WITWER K W, AIKAWA E, et al. Minimal information for studies of extracellular vesicles 2018 (MISEV2018): a position statement of the International Society for Extracellular Vesicles and update of the MISEV2014 guidelines [J]. Journal of extracellular vesicles, 2018, 7: 1535750.

[29] THÉRY C, AMIGORENA S, RAPOSO G, et al. Isolation and characterization of exosomes from cell culture supernatants and biological fluids [J]. Current protocols in cell biology, 2006, Chapter 3: p. Unit 3.22.

[30] KOWAL J, TKACH M, THÉRY C. Biogenesis and secretion of exosomes [J]. Current opinion in cell biology, 2014, 29: 116 – 125.

[31] TODOROVA D, SIMONCINI S, LACROIX R, et al. Extracellular vesicles in angiogenesis [J]. Circulation research, 2017, 120: 1658 – 1673.

[32] EL ANDALOUSSI S, MÄGER I, BREAKEFIELD X O, et al. Extracellular vesicles: biology and emerging therapeutic opportunities [J]. Nature reviews drug discovery, 2013, 12: 347 –357.

[33] GOULD S J, RAPOSO G. As we wait: coping with an imperfect nomenclature for extracellular vesicles [J]. Journal of extracellular vesicles, 2013, 2: 20389.

[34] KIM D K, KANG B, KIM O Y, et al. EVpedia: an integrated database of high-

throughput data for systemic analyses of extracellular vesicles [J]. Journal of extracellular vesicles, 2013, 2: 10.3402.

[35] KALRA H, SIMPSON R J, JI H, et al. Vesiclepedia: a compendium for extracellular vesicles with continuous community annotation [J]. PLOS biology, 2012, 10: e1001450.

[36] SIMPSON R J, KALRA H, MATHIVANAN S. ExoCarta as a resource for exosomal research [J]. Journal of extracellular vesicles, 2012, 1: 10.3402.

[37] ZABOROWSKI M P, BALAJ L, BREAKEFIELD X O, et al. Extracellular vesicles: composition, biological relevance, and methods of study [J]. Bioscience, 2015, 65: 783 – 797.

[38] RAPOSO G, STOORVOGEL W. Extracellular vesicles: exosomes, microvesicles, and friends [J]. Journal of cell biology, 2013, 200: 373 – 383.

[39] LARSON M C, WOODLIFF J E, HILLERY C A, et al. Phosphatidylethanolamine is externalized at the surface of microparticles [J]. Biochimica et biophysica acta, 2012, 1821: 1501 – 1507.

[40] LAULAGNIER K, MOTTA C, HAMDI S, et al. Mast cell- and dendritic cell-derived exosomes display a specific lipid composition and an unusual membrane organization [J]. Biochemical Journal, 2004, 380: 161 – 171.

[41] RATAJCZAK J, WYSOCZYNSKI M, HAYEK F, et al. Membrane-derived microvesicles: important and underappreciated mediators of cell-to-cell communication [J]. Leukemia, 2006, 20: 1487 – 1495.

[42] CHEN T S, LAI R C, LEE M M, et al. Mesenchymal stem cell secretes microparticles enriched in pre-microRNAs [J]. Nucleic acids research, 2010, 38: 215 – 224.

[43] BATRAKOVA E V, KIM M S. Development and regulation of exosome-based therapy products [J]. Wiley interdisciplinary reviews-nanomedicine and nanobiotechnology, 2016, 8: 744 – 757.

[44] LIMA L G, CHAMMAS R, MONTEIRO R Q, et al. Tumor-derived microvesicles modulate the establishment of metastatic melanoma in a phosphatidylserine-dependent manner [J]. Cancer letters, 2009, 283: 168 – 175.

[45] CAI Z, YANG F, YU L, et al. Activated T cell exosomes promote tumor invasion via Fas signaling pathway [J]. Journal of immunology, 2012, 188: 5954 – 5961.

[46] PISITKUN T, SHEN R F, KNEPPER M A. Identification and proteomic profiling of exosomes in human urine [J]. Proceedings of the National Academy of Sciences of the United States of America, 2004, 101: 13368 – 13373.

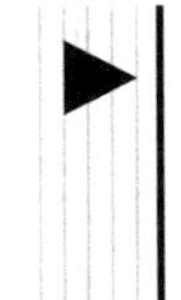

Chapter 9 Genome Editing and Gene Therapy
第九章 基因编辑与基因治疗

[**中文导读**]

基因编辑是一种新兴的比较精确的能对生物体基因组特定目标基因进行修饰的基因工程技术。基因编辑依赖于经过基因工程改造的核酸酶，也称“分子剪刀”，在基因组中特定位置产生位点特异性双链断裂，诱导生物体通过非同源末端连接或同源重组来修复双链断裂，但这个修复过程容易出错，从而导致靶向突变。这种靶向突变就是基因编辑。

基因治疗是指通过给细胞注入 DNA（或 RNA）来修饰基因表达或纠正突变/缺陷基因，以治疗疾病的一套策略。经典的基因治疗只能将外源基因添加到细胞中，通过过表达基因产物达到治疗的效果。基因编辑技术则更加灵活，可以通过定点整合来修复突变基因或者过表达治疗性基因产物，可以删除基因编码区部分序列达到基因失活的效果，还能作用于基因的启动子序列来增强或者减弱基因表达，故基因编辑技术有望显著提高基因治疗的持久性和安全性。以锌指核酸酶（zinc fingernuclease，ZFN）、转录激活因子样效应物核酸酶（transcription activator-like effector nuclease，TALEN）、规律成簇的间隔短回文重复序列/Cas 核酸酶（clustered regularly interspaced short palindromic repeats/CRISPR-associated systems，CRISPR/Cas）为代表的基因编辑技术蓬勃发展，为细胞治疗领域带来了开创性突破。基因治疗作为一种新型的生物技术疗法，为解决许多疾病提供了新的思路和契机。

1 Human Genome and Human Diseases

1.1 Introduction to Genes and Diseases

Genes are DNA fragments with genetic effects. They are the units of heredity (genetic factors) that control biological traits. They are specific nucleotide sequences with genetic information on DNA or RNA molecules. Genes transmit genetic information to the next generation through duplication, so that the offspring will appear in similar traits to the parent. It also changes this self-associative property through mutations, storing all the information about the process of life gestation, growth, and apoptosis. Through replication, transcription, and expression, it completes important physiological processes such as life reproduction, cell division, and protein synthesis.

Among human diseases, diseases determined by genetic factors or mainly genetic factors are called genetic diseases. According to clinical statistics, 25% of physical defects, 30% of childhood diseases and 60% of adult diseases are caused by genetic diseases. As for human genetic diseases, it is reported that there are 5 000 kinds, most of which are caused by single gene defects. The body is a complex dynamic balance system. The influence of each gene on the normal function of the body is complex, and the change of any gene will cause a variety of symptoms. Because the causes of these diseases are complex and occur at the level of genetic material, it is difficult to achieve the goal of radical cure with traditional treatment methods.

Monogenic diseases are caused by the presence of disease-causing genes at a certain locus (Locus) in the genome, such as: thalassemia[1], hemophilia[2], sickle-cell anemia[3], etc. Others are caused by the synergistic effects of faulty genes in multiple gene loci. In many cases, these defective genes also require the participation of certain environmental factors to cause individual disease. This type of disease is called polygenic disease, such as hypertension, diabetes, coronary heart disease, tumors, and so on.

The occurrence of genetic diseases is often caused by gene deletion, mutation, dislocation or insertion of foreign genes (or DNA fragments), and the direct cause of disease symptoms is often the changes in the products of gene control. Changes in the primary, secondary, tertiary, or quaternary structure of gene products (proteins, enzymes) can cause diseases.

1.2 Monogenic Diseases

Monogenic diseases result from modifications in a single gene occurring in all cells of the body. Though relatively rare, they affect millions of people worldwide. Scientists currently estimate that over 10 000 types of human diseases are known to be monogenic. Pure genetic diseases are caused by a single error in a single gene in the human DNA. The nature of disease depends on the functions performed by the modified gene. The single-gene or monogenic diseases can be classified into three main categories: recessive, dominant, and X-linked.

All human beings have two sets or copies of each gene called "allele", one copy on each side of the chromosome pair. Recessive diseases are monogenic diseases that occur due to damages in both copies or allele. Dominant diseases are monogenic disorders that involve damage to only one gene copy. X linked diseases are monogenic disorders that are linked to defective genes on the X chromosome which is the sex chromosome. The X linked alleles can also be dominant or recessive. These alleles are expressed equally in men and women, more so in men as they carry only one copy of X chromosome (XY) whereas women carry two (XX).

Monogenic diseases are responsible for a heavy loss of life. The global prevalence of all single gene diseases at birth is approximately 10/1000.

1.2.1 Thalassaemia

Thalassaemia is a blood related genetic disorder which involves the absence of/errors in genes responsible for production of haemoglobin, a protein present in the red blood cells. Each red blood cell can contain 240 – 300 million molecules of haemoglobin. The severity of the disease depends on the mutations involved in the genes, and their interplay.

A haemoglobin molecule has subunits commonly referred to as α and β. Both subunits are necessary to bind oxygen in the lungs properly and deliver it to tissues in other parts of the body. Genes on chromosome 16 are responsible for α subunits, while genes on chromosome 11 control the production of β subunits. A lack of a particular subunit determines the type of thalassaemia (e. g. a lack of α subunits results in α-thalassemia). The lack of subunits thus corresponds to errors in the genes on the appropriate chromosomes.

The α and β thalassaemia are the most common inherited single-gene disorders in the world with the highest prevalence in areas where malaria was or still is endemic.

1.2.2 Haemophilia

Haemophilia is a hereditary bleeding disorder, in which there is a partial or total lack of an essential blood clotting factor. It is a lifelong disorder, which results in excessive bleeding, and many times spontaneous bleeding, which, very often, is internal. Haemophilia A, the most common form, referred to as classical haemophilia, is the result of a deficiency in clotting factor 8, while haemophilia B (Christmas disease) is the result of clotting factor 9 deficiency. This illness is a sex-linked recessive disorder.

Due to the sex-linkage of the disorder, there is a greater prominence in males than in females. About a third of new diagnoses are where there is no previous family history. It appears world-wide and occurs in all racial groups.

1.2.3 Sickle Cell Anemia

Sickle cell anemia is a blood related disorder that affects the haemoglobin molecule, and causes the entire blood cell to change shape under stressed conditions. In sickle cell anaemia, the haemoglobin molecule is defective. After haemoglobin molecules give up their oxygen, some may cluster together and form long, rod-like structures which become stiff and assume sickle shape.

Unlike healthy red blood cells, which are usually smooth and donut-shape, sickled red blood cells cannot squeeze through small blood vessels. Instead, they stack up and cause blockages that deprive organs and tissues of oxygen-carrying blood. This process produces periodic episodes of pain and ultimately can damage tissues and vital organs and lead to other serious medical problems. Normal red blood cells live about 120 days in the bloodstream, but sickled red cells die after 10 – 20 days. Because they cannot be replaced fast enough, the blood is chronically short of red blood cells, leading to a condition commonly referred to as anemia.

Sickle cell anemia affects millions of people throughout the world. It is particularly common among people whose ancestors come from Sub-Saharan Africa, South America, Cuba, Central America, Saudi Arabia, India, and Mediterranean countries such as Turkey, Greece, and Italy. In the Unites States, it affects around 72 000 people, most of whose ancestors come from Africa. The disease occurs in about 1 in every 500 African-American births and 1 in every 1 000 to 1 400 Hispanic-American births. About 2 million Americans, or 1 in 12 African Americans, carry the sickle cell allele.

1.3 Polygenic Diseases

Polygenic disease (or polygenic disorder) results from the effects of the combined action or interaction of multiple genes. More than one gene is associated with a polygenic disease or disorder. Because of the complexity of genetic interactions, it does not follow a pattern the same way a monogenic disease does. In humans, though, polygenic conditions occur more frequently than monogenic diseases.

Apart from the multiple genes involved, non-genetic factors may also play a role in the manifestation of the disease. For instance, type 2 diabetes is a disease characterized by the presence of excessive glucose levels circulating in blood and symptoms such as polyphagia, polyuria, and polydipsia. It is a polygenic disease having involved with more than one gene, i. e. more than 36 genes (e. g. TCF7L2 allele). These genes increase the risk of developing diabetes. Many of these genes are linked to beta cell functions. Other factors associated with the contraction of diabetes include non-genetic factors (e. g. lifestyle and diet).

Other examples of polygenic conditions are coronary heart disease, hypertension, autoimmune diseases, cancers, obesity, and atherosclerosis.

1.4 Acquired Genetic Diseases

Acquired genetic disease refers to the genetic alteration or loss of host cells caused by pathogenic microorganism infection, but it is not inherited. For example, HBV and HCV can be integrated into the DNA of liver cells to cause liver cancer.

2 The Development History and Classification of Gene Therapy

2.1 Definition of Gene Therapy

Gene therapy is the insertion of genes into an individual's cells and tissues to treat a disease, and hereditary diseases in which a defective mutant allele is replaced with a functional one. Gene therapy is a technique that modifies a person's genes to treat or cure disease[4].

2.2 The Development Process of Gene Therapy in the Past 30 Years

The concept of gene therapy was proposed in the early 1970s[5]. In the early stages of gene therapy, gene addition is a major challenge. Initially, until the invention of viral vectors in the 1980s, there were no effective tools to deliver foreign genes to human cells. [6-8] Gene therapy clinical research began in 1990, when the National Institutes of Health conducted the first clinical research for a rare immunodeficiency disease. In the 1990s, the four-year-old A-shanti was severely ill with combined immunodeficiency (SCID), a rare genetic disease in which the immune system almost collapsed due to a genetic defect in adenosine deaminase. Dr. Anderson of the United States and other collaborators extracted white blood cells from Ashanti, introduced normal adenosine deaminase into these cells, and then re-infused the genetically modified white blood cells into the patient's body. After the operation, Ashanti produced normal adenosine deaminase[9].

The successful case of Ashanti ignited people's enthusiasm for gene therapy. After that, many scientists in many countries quickly invested in the wave of gene therapy. Many people believed that the era of gene therapy had come. Until 1999, 18-year-old Jessie died unfortunately in a gene therapy clinical trial due to a severe immune response caused by a viral vector, which sounded the alarm for scientists and asked people to re-examine the gene therapy technology[10]. Since then, gene therapy has failed in several clinical trials. In 2000, treatment of X-linked severe combined immunodeficiency (SCID-X1) with a γ-retroviral vector resulted in T-cell leukemia in nearly half of patients[11-13]. These failures brought gene therapy to a low ebb, but also provided valuable lessons for immune response and insertional mutagenesis. Until recent years, the safety and effectiveness of viral vectors have been continuously improved, and the rise of gene editing technology has caused gene therapy to become increasingly popular.

The first milestone was that Glybera was approved in Europe in 2012. This is the first gene therapy product in Western countries and the first gene therapy targeting genetic diseases around the world. Although Glylbera was not commercially successful, it opened the door to gene therapy[14]. In the next few years, several gene therapy products flooded the market. Strimvelis was approved in Europe in 2016[15]. The year 2017 is considered to be a breakthrough year for gene therapy. In this year, two CAR-T therapies (Kymriah[16] and Yescarta[17]) and Luxturna[18] were approved by the U. S. Food and Drug Administration.

2.3 Advances in Gene Editing Technology Have Given a Boost to Gene Therapy

The breakthrough of gene editing technology also promotes the development of gene therapy to a certain extent. The artificial endonuclease-mediated genome editing technology mainly includes three types: ZFNs, TALENs, and CRISPR/Cas9 technologies. Compared with virus vectors in traditional genetic engineering, gene editing technology provides a precise "scalpel" to increase, decrease, and modify genes.

2.4 Gene Therapy Strategies

The term gene therapy is a broad term: it covers many different strategies, all of which are aimed at overcoming or alleviating the disease by introducing genes, gene fragments or oligonucleotides into the process of the affected individual's cells.

2. 4. 1 Gene Correction

Gene correction refers to correcting the abnormal bases in the disease-causing genes, while retaining the normal parts.

2. 4. 2 Gene Replacement

Gene replacement refers to the use of normal genes through homologous recombination technology to replace disease-causing genes in situ, so that the DNA in the cell can be completely restored to a normal state.

2. 4. 3 Gene Supplement

Gene supplement refers to the introduction of normal genes into somatic cells and their expression through non-specific integration of genes to compensate for the function of defective genes, or to enhance the function of original genes, but the pathogenic genes themselves are not removed.

2. 4. 4 Gene Inactivation

Gene inactivation refers to the introduction of specific antisense nucleic acids (antisense RNA and antisense DNA) and ribozymes into cells to block the abnormal expression of certain genes at the transcription and translation levels to achieve the purpose of treatment.

2. 4. 5 Suicide Gene

Suicide gene refers to a gene in some viruses or bacteria that can produce an enzyme, which can convert the original non-cytotoxic or low-toxic drug precursors into cytotoxic substances and kill the cells themselves.

2. 4. 6 Immunity Therapy

Immune gene therapy is the introduction of genes corresponding to antiviral or tumor immunity and antigenic determinants into body cells to achieve the purpose of treatment, such as the introduction and expression of cytokine genes.

2. 4. 7 Resistance Treatment

Drug-resistant gene therapy is to improve the body's ability to tolerate chemotherapeutic drugs during tumor treatment. Genes that produce drug-resistant toxicity are introduced into human cells so that the body can tolerate larger doses of chemotherapy, such as the introduction of mdr-1 in the multidrug resistance gene into bone marrow stem cells.

2. 5 Classification of Gene Therapy

In a broad sense, gene therapy refers to the transfer of a certain genetic material into the patient's body and normal expression, which has achieved the goal of treating related diseases. Gene therapy in a narrow sense refers to correcting or replacing the original defective genes with genes that perform normal functions to achieve the purpose of treatment.

At present, there are roughly three classification methods for gene therapy, including classification according to route of administration, operation methods, and target cells.

According to the route of administration, it can be divided into *in vivo* and *in vitro* gene therapy[19].

2.5.1 *In Vivo* Approach

This is to assemble foreign genes into specific eukaryotic expression vectors and directly introduce them into the body. This vector can be viral or non-viral, or even naked DNA. The *in vivo* gene transfer route is simple to operate and easy to promote, but it is not yet mature. There are a series of problems such as short duration of efficacy, immune rejection, and safety.

2.5.2 *In Vitro* Approach

This refers to the introduction of vectors containing foreign genes into the body's own or allogeneic cells (or heterogeneous cells) *in vitro*, and then transfer back to the human body after *in vitro* cell amplification. The *in vitro* gene transfer method is relatively classic, safe, and the effect is easier to control, but it is not easy to promote because of the many steps, complex technology, and difficulty.

According to the operation mode, it can be divided into modified replacement and supplementary gene therapy.

(1) Gene correction and gene replacement. Gene correction and gene replacement refer to the correction of the abnormal sequence of the defective gene, and the accurate repair of the defective gene in situ, without any other changes in the genome. Through homologous recombination (gene targetting) technology, the foreign normal genes are recombined in specific parts, so that the defective genes can be specifically repaired in situ.

(2) Gene augmentation and gene inactivation. Gene augmentation and gene inactivation refer to the expression of normal products by introducing foreign genes without removing abnormal genes, thereby compensating for the functions of defective genes; or specifically blocking the translation or transcription of certain genes to achieve the suppression of certain abnormal gene expression.

Whether it is gene correction replacement or supplemental inactivation, it is required to be able to clearly identify each gene that cures diseases, to have a detailed study and understanding of the target gene and its products, and to ensure that the introduced gene can exist stably in the target cell for a long time. To play a role in moderate expression, it is also necessary to ensure that the method and vector of gene introduction are safe and harmless to the host cell.

According to different target cells, it can be divided into somatic cell and germ-line cell gene therapy.

(1) Somatic cell gene therapy. Somatic cells such as HSCs, skin fibroblasts, hepato-

cytes, vascular endothelial cells, lymphocytes, muscle cells and tumor cells, etc. transfer normal genes to certain types of target cells mentioned above, which are generally expanded and cultured through *in vitro* methods. It is then transferred back into the body to achieve the purpose of gene therapy of diseases.

(2) Germ-line cell gene therapy. Germ-line cell gene therapy is the introduction of normal genes into sperm, egg cells or early embryonic cells in the patient's body to express related products to treat diseases.

The biggest difference between somatic cell and germ-line cell gene therapy is whether it will affect the offspring. Germ-line cell gene therapy not only affects the present, but also inherits genes to future generations. Because the current gene therapy technology is not yet matures and involves a series of ethical issues, germ-line cell gene therapy is still a forbidden area. Under existing conditions, gene therapy is limited to somatic cells.

2.6 The Main Application of Gene Therapy in Cancer

At present, the application of gene therapy in the field of tumors is mainly through genetic modification of T cells *in vitro*, in the form of CAR-T (chimeric antigen receptor T cell) or TCR-T (T cell receptor-gene-engineered T cells), and then infusion of genetically modified T cells into the patient's body, so that the modified T cells attack cancer cells to achieve the purpose of treating tumors.

2.6.1 CAR-T: Popular Technology for Tumor Immunotherapy

CAR-T, simply put, is a chimeric antigen receptor which is mainly composed of three parts: extracellular antigen binding region, the transmembrane linking region, and the intracellular signal region added to T cells (an important immune cell in the human body) through genetic modification technology[20], so that the immune T cells not only to specifically recognize cancer cells, but also to activate T cells to kill cancer cells[21].

The process of CAR-T technology to treat cancer mainly consists of four steps: ① Collection and activation of T cells; ② *In vitro* gene transduction of T cells: through genetic engineering technology, let CAR be chimerized onto T cells; ③ *In vitro* proliferation culture of constructed CAR-T cells; ④ CAR-T reinfusion and patient observation.

2.6.2 TCR-T: Tumor Immunotherapy That Is Expected to Break Through in the Field of Solid Tumors

TCR-T refers to the direct modification of the surface receptor of T cells to recognize tumor antigens: T cell receptor (TCR) by means of genetic engineering, thereby strengthening the ability of T cells to recognize and kill tumor cells.

Although, like CAR-T therapy, it belongs to tumor immunotherapy, it is also a gene therapy for genetic modification of T cells *in vitro*. However, the difference from CAR-T therapy is: ① CAR-T therapy is to construct a chimeric antibody receptor on T cells, while

TCR-T technology is to modify the surface receptor of T cells to enhance their affinity; ② Unlike CAR-T which can only recognize tumor cell surface antigens, TCR-T can target tumor antigens in the cell or on the cell surface, so that TCR-T has more targets to choose from, especially including multiple intracellular antigens related to cancer, which is why some scho-lars believe that TCR-T has great potential in the field of solid tumors. For example, cancer-testis antigen (CTAs) is a type of cytoplasmic protein that is found in a variety of tumor tissues but is rarely found in normal cells.

Of course, there are still many challenges for TCR-T therapy. No product has been approved yet. The difficulties include: inhibition of tumor microenvironment, off-target toxicity, neurotoxicity, and cytokine release syndrome (CRS) similar to that encountered by CAR-T therapy and so on.

2.7 Application of Gene Therapy in the Field of Rare Diseases

There are about 7 000 rare diseases confirmed globally, but only a few hundred rare diseases have approved therapeutic drugs. Gene therapy is of great significance in the field of rare diseases, because more than 80% of rare diseases are caused by single gene defects. For rare diseases, traditional small molecule drugs usually work by reducing symptoms. On the contrary, gene therapy has the potential to correct gene defects, especially for single-gene rare diseases, providing a potential cure instead of simple management symptom. Furthermore, successful gene therapy may only require one treatment, rather than a lifelong continuous treatment.

2.7.1 Application of Gene Therapy in Ophthalmology

Inherited retinal diseases (IRDs) are a group of rare eye diseases caused by genetic mutations that often lead to vision loss or blindness. IRDs can be further subdivided into retinitis pigmentosa (RP), leber congenital amaurosis (LCA), congenital stationary night blindness (CSNB) and so on. Because IRDs are mostly caused by a single gene defect (the incidence of IRDs caused by a single gene defect is about 1/3 000), and because the eye is a relatively independent organ in the human body that is easy to treat, IRDs have become a hot topic in gene therapy research areas.

Luxturna, which was approved for listing by Spark in 2017, is a type of hereditary retinopathy used to treat RPE65 gene defects, requiring only one-time treatment. Luxturna carries the normal RPE65 gene through the adenovirus-associated virus (AAV) and enters the retina. The normal RPE65 gene is not integrated into the DNA of human cells, but the normal RPE65 protein is synthesized in the nucleus to help trigger the light transmission path restores normal visual function[22-24].

2.7.2 Application of Gene Therapy in the Field of Hemophilia

Hemophilia mainly includes hemophilia A and hemophilia B, both of which are rare X chromosome recessive genetic diseases. Due to the lack of sufficient coagulation factor Ⅷ and coagulation factor Ⅸ proteins, respectively, the coagulation function is abnormal and persistent bleeding can affect life in severe cases. The Global Hemophilia Alliance predicts that more than 150 000 people worldwide have hemophilia A, and nearly 30 000 people have hemophilia B.

BioMarin's BMN270 targets hemophilia A, and like Spark, it also uses the AAV[25]. At the 17th American ASH Conference, BioMarin announced the latest data of BMN270, showing that for seven patients with an injection dose of 6.0×10^{13} μg/kg, after 78 weeks of treatment, the median and average levels of coagulation factor Ⅷ were respectively reaching 90% and 89%, according to the definition of the Global Hemophilia Alliance, the level of factor Ⅷ in normal people is between 50% and 150%. At the same time, the data shows that after four weeks of treatment, the median annualized bleeding rate and the number of injections of coagulation factor Ⅷ are both 0, which means that there is no bleeding phenomenon and receiving traditional treatment.

2.7.3 Application of Gene Therapy in Neurogenetic Diseases

For a long time, gene therapy has been very challenging in the treatment of neurological diseases. The difficulties include the safety of viral vectors and the way of delivering viral vectors. Today, tremendous progress has been made in the above two aspects. In particular, scientists have discovered that the relative safety and effectiveness of the AAV9 viral vector, as well as its ability to cross the blood-brain barrier, has become the carrier of choice for multiple neurological disease gene therapy projects.

AVXS-101 of AveXis is a gene therapy product used to treat spinal muscular atrophy (SMA). SMA is a serious neuromuscular disease caused by a defect in the SMN1 gene, with an incidence of approximately one in every 10 000 newborn children[26]. In the AVXS-101 gene therapy product, the AAV9 virus vector is used to carry the normal working human SMN gene and is administered intravenously[27]. The SMN gene carried by the virus into the cell body is not integrated into the original DNA molecule of the cell. But through transcription and translation, the SMN protein is produced at the appropriate level.

2.7.4 Application of Gene Therapy in Other Rare Genetic Diseases

The rare diseases currently under study in gene therapy also include sickle-type anemia, β-thalassemia, lipoproteinase deficiency, mucopolysaccharidosis, adrenal leukodystrophy, etc.

2.8 The Problems and Challenges of Gene Therapy

It has only been a few years since gene therapy has been applied to the clinic. Although it has promoted the development of basic medicine and clinical medicine, we should be soberly aware that there are still many problems in current gene therapy.

2.8.1 Problems with Gene Therapy Itself

Although the development of modern science and technology has made gene therapy more and more perfect, there are still problems such as fewer target genes, lack of efficient vector systems, and poor controllability of target gene expression. Therefore, to study the pathogenesis of diseases, better understand pathogenic gene and their mechanism of action, improve the efficiency, targeting and safety of gene transfer, in-depth study gene expression and regulation, establish good experimental animal models and provide more reliable data are the problem that should be addressed in the current gene therapy research.

2.8.2 Ethical Issues

Gene therapy is expensive and not available to everyone. This is contrary to the nature of medicine as a social welfare undertaking. In addition, as far as gene therapy of germ-line cell is concerned, gene therapy will change the DNA sequence in sex cells and pass this change to offspring. At the same time, gene therapy will increase the degree of aging, which will increase the burden on society. In addition, gene therapy also involves patients' right to know and privacy issues.

2.8.3 Security Issues

The safety of gene therapy means that gene therapy should ensure that the introduction of exogenous gene should not cause new harmful genetic mutations. Therefore, it is quite critical to choose a safe and effective vector when performing gene therapy. The vector should not cause insertion mutations which may cause the inactivation of an important gene or activate a proto-oncogene, with disastrous consequences. The exogenous gene introduced at the same time generally do not have an expression control system, and their expression level may affect some physiological activities of the body.

2.9 Prospects for Gene Therapy

As a brand-new disease treatment method, gene therapy has achieved encouraging results since its birth more than 20 years ago. With the development of related disciplines such as molecular pathology and molecular biology, treatment strategies have become increasingly abundant, many of which have entered clinical trials or implementation stages and have become clinically important adjuvant treatments. At the same time, the diseases targeted for treatment have also expanded from early single-gene genetic diseases to tumors, viral diseases, cardiovascular diseases and other diseases that seriously threaten human health. At present, gene therapy still faces difficulties in theory and technology, such as fewer therapeutic genes, low gene transfer efficiency, and poor controllability of gene expression. There are also many controversies in ethics and morality, but it is foreseeable that after overcoming these challenges, gene therapy will truly become a conventional treatment method in the near future and make an important contribution to the maintenance of human health.

3 Genome Editing

3.1 Definition of Genome Editing

Genome editing is a technique for precisely and effectively modifying DNA in cells[19]. It uses enzymes called "engineered nucleases" to cut specific sequences of DNA[28]. It can be used to add, delete or change the DNA in the genome so that the characteristics of cells or organisms can be changed[29].

3.2 Application of Genome Editing

Genome editing is an attractive and challenging therapeutic approach that can correct or eliminate mutations that lead to the development of cancer and other gene-driven diseases. So far, *in vitro* genome editing, or *in vitro* genetic engineering of cell *in vitro*, has been the most widely used approach, in which modified cells are re-implanted into patients.

3.2.1 Genome Editing for Disease Modeling Research

Disease animal models have always been an important resource for advancing the field of biomedicine. With the help of genome editing technology, many applicable models with specific mutations have been produced, which can mimic clinical phenotypes (Fig. 9-1). Disease models are usually constructed by introducing site-specific modifications in ESCs. In addition, diseased animals can be generated by editing induced mutant zygotes or editing somatic cells combined with somatic cell nuclear transfer (SCNT) technology. In addition, the engineered virus loaded with editing elements can be applied to target diseases to be destroyed. After the virus is injected into adult animals, the tissue-specific expression of editing elements provides a tool to modify genes quickly and accurately, thereby avoiding embryonic lethality[30].

3.2.2 Applications of Genome Editing Technologies in Gene Therapy

Genome editing technologies are not only used to generate disease animal models, but also destined to enter the field of therapy. There are many treatment methods based on genome editing: inactivate or correct harmful mutations, introduce protective mutations, insert therapeutic exogenous genes, and destroy viral DNA. Many proof-of-principle studies have shown examples of successful gene therapy depending on genome editing. Obtaining therapeutic modifications requires delivery of engineered nucleases to target cells, which can be achieved *in vitro* or *in vivo*[30].

3.2.3 Applications of Genome Editing in Biotechnology

Genome editing has been used in agriculture to genetically modify crops to improve their yields and resistance to disease and drought, as well as to genetically modify cattle that do not have horns[31-32].

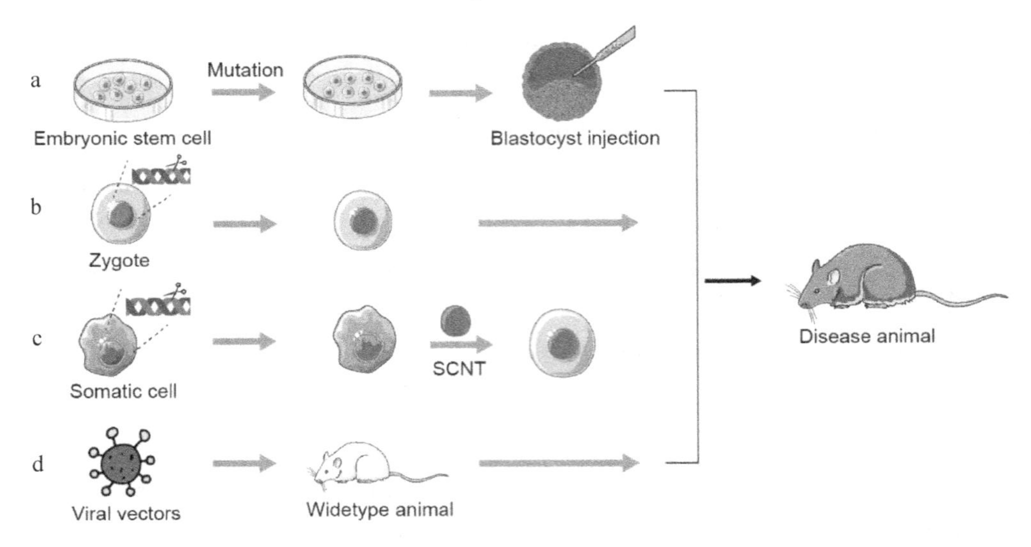

a. Cultured ESCs can be used to introduce morbigenous mutations using genome editing tools. The edited ESCs can be injected into host blastocysts, whereafter are implanted into pseudo-pregnant to produce disease animal. b. Animal zygote is directly edited and the edited zygote is developed into diseased model. c. Disease animal could be generated by combining somatic cell genome editing and SCNT technology. d. Genome editing elements are packaged by viral vectors. Disease animal can be generated by administration of engineered virus.

Fig. 9 – 1 Schematic overview of constructing disease animal model in four major ways

Illustrated by 苏丹丹.

References: LI Q, QIN Z, WANG Q, et al. Applications of genome editing technology in animal disease modeling and gene therapy [J]. Computational and structural biotechnology journal, 2019, 17: 689 – 698.

3.3 Genome Editing 1.0: ZFN Tool

ZFNs are artificial restriction enzymes produced by fusing a zinc finger DNA binding domain with a DNA cleavage domain. Zinc finger domains can be engineered to target specific desired DNA sequences, which enables ZFNs to target unique sequences in complex genomes. By using endogenous DNA repair mechanisms, these reagents can be used to precisely alter the genomes of higher organisms[29, 33].

3.3.1 ZFN Structure and Basic Technical Principles

ZFN is a site-specific endonuclease designed to bind and cleave DNA at a specific location. There are two protein domains. The first domain is the DNA binding domain, which is composed of eukaryotic transcription factors and contains zinc fingers. The second domain is the nuclease domain, which consists of the FokI restriction enzyme and is responsible for the catalytic cleavage of DNA.

The DNA-binding domains of individual ZFNs typically contain between three and six individual zinc finger repeats and can each recognize between 9 and 18 basepairs[34]. If the zinc finger domains are perfectly specific for their intended target site then even a pair of

3-finger ZFNs that recognize a total of 18 basepairs can, in theory, target a single locus in a mammalian genome. The most straightforward method to generate new zinc-finger arrays is to combine smaller zinc-finger "modules" of known specificity. The most common modular assembly process involves combining three separate zinc fingers that can each recognize a 3 basepair DNA sequence to generate a 3-finger array that can recognize a 9 basepair target site.

The non-specific cleavage domain from the type Ⅱ restriction endonuclease FokI is typically used as the cleavage domain in ZFNs[35]. This cleavage domain must dimerize in order to cleave DNA and thus a pair of ZFNs are required to target non-palindromic DNA sites. Standard ZFNs fuse the cleavage domain to the C-terminus of each zinc finger domain. In order to allow the two cleavage domains to dimerize and cleave DNA, the two individual ZFNs must bind opposite strands of DNA with their C-termini a certain distance apart.

3.3.2 Application of ZFN Technology

ZFNs are useful to manipulate the genomes of many plants and animals. ZFNs are also used to create a new generation of genetic disease models called isogenic human disease models. ZFNs have also been used in a clinical trial of $CD4^+$ human T cells with the CCR5 gene disrupted by ZFNs to be saved as a potential treatment for HIV/AIDS[36-38]. Custom-designed ZFNs that combine the non-specific cleavage domain (N) of FokI endonuclease with zinc-finger proteins (ZFPs) offer a general way to deliver a site-specific DSB to the genome, and stimulate local homologous recombination by several orders of magnitude. This makes targeted gene correction or genome editing a viable option in human cells. Since ZFN-encoding plasmids could be used to transiently express ZFNs to target a double strand break (DSB) to a specific gene locus in human cells, they offer an excellent way for targeted delivery of the therapeutic genes to a pre-selected chromosomal site. The ZFN-encoding plasmid-based approach has the potential to circumvent all the problems associated with the viral delivery of therapeutic genes. The first therapeutic applications of ZFNs are likely to involve *in vitro* therapy using a patient own stem cells[37]. After editing the stem cell genome, the cells could be expanded in culture and reinserted into the patient to produce differentiated cells with corrected functions. The initial targets is likely to include the causes of monogenic diseases such as the IL2Rγ gene and the β-globin gene for gene correction and CCR5 gene for mutagenesis and disablement.

3.3.3 Defects of ZFN Technology

Although ZFN technology is simple and practical, it also has certain drawbacks. The cleavage of DNA by ZFN requires the dimerization of two FokI cleavage regions and requires at least one recognition unit to bind DNA. Although the DNA recognition domain has strong specific recognition ability, since the process of ZFN shearing does not completely depend on the formation of homodimers, once a heterodimer is formed, it is likely to cause off-target effects and eventually may cause DNA mismatch and sequence changes, resulting in strong cytotoxicity. When these adverse effects accumulate too much and exceed the range that the cell repair mechanism can withstand, it will cause cell apoptosis. On the other hand, this method is still limited by the existing research methods in the field of biology, so the accuracy and consequences of the operation inside the cell are more difficult to predict. If ZFN causes mutations in related genes, it may lead to a series of unexpected consequences. In applications related to the human body, it may even cause cancer. In addition, ZFN, as one of the means of gene therapy, may trigger an immune response if used in organisms. Existing research methods cannot predict whether the introduced ZFN protein will cause an attack on the immune system. And so far, ZFN technology can only be used for *in vitro* operations (*in vitro*), after the cells extracted from the human body are processed, they are introduced back into the patient's body. However, directly introducing relevant ZFN elements into the patient's body for gene editing has a greater potential risk, and the efficiency is not high. Many of the above restrictions make the ZFN operations related to the human body more cumbersome and difficult to promote and apply.

3.4 Genome Editing 2.0: TALEN Tool

TALEN technology leverages artificial restriction enzymes generated by fusing a TAL effector DNA-binding domain to a DNA cleavage domain. Restriction enzymes are enzymes that cut DNA strands at a specific sequence. Transcription activator-like effectors (TALEs) can be quickly engineered to bind practically any desired DNA sequence. By combining such an engineered TALE with a DNA cleavage domain (which cuts DNA strands), one can engineer restriction enzymes that will specifically cut any desired DNA sequence. When these restriction enzymes are introduced into cells, they can be used for gene editing or for genome editing in situ, a technique known as genome editing with engineered nucleases[39].

3.4.1 TALEN Structure

TALEs are proteins that are secreted by Xanthomonas bacteria. The DNA binding domain contains a repeated highly conserved 33 – 34 amino acid sequence with divergent 12th and 13th amino acids[40]. These two positions, referred to as the repeat variable diresidue (RVD), are highly variable and show a strong correlation with specific nucleotide recognition[41]. This relationship between amino acid sequence and DNA recognition has allowed for the engineering of specific DNA-binding domains by selecting a combination of repeated segments containing the appropriate RVDs[42].

The non-specific DNA cleavage domain from the end of the FokI endonuclease can be used to construct hybrid nucleases that are active in many different cell types. The FokI domain functions as a dimer, requiring two constructs with unique DNA binding domains for sites in the target genome with proper orientation and spacing. Both the number of amino acid residues between the TALE DNA binding domain and the FokI cleavage domain and the number of bases between the two individual TALEN binding sites appear to be important parameters for achieving high levels of activity.

3.4.2 TALEN Mechanism

The simple relationship between amino acid sequence and DNA recognition of the TALE binding domain allows for the efficient engineering of proteins. Once the TALEN constructs have been assembled, they are inserted into plasmids; the target cells are then transfected with the plasmids, and the gene products are expressed and enter the nucleus to access the genome. Alternatively, TALEN constructs can be delivered to the cells as mRNAs, which removes the possibility of genomic integration of the TALEN-expressing protein. Using an mRNA vector can also dramatically increase the level of homology directed repair (HDR) and the success of introgression during gene editing.

TALEN technology can be used to edit genomes by inducing DSB, which cells respond to with repair mechanisms. Non-homologous end joining (NHEJ) reconnects DNA from either side of a DSB where there is very little or no sequence overlap for annealing[43]. This repair mechanism induces errors in the genome via insertion or deletion, or chromosomal rearrangement; any such errors may render the gene products coded at that location non-functional. Because this activity can vary depending on the species, cell type, target gene, and nuclease used, which should be monitored when designing new systems. Alternatively, DNA can be introduced into a genome through NHEJ in the presence of exogenous double-stranded DNA fragments. Homology directed repair can also introduce foreign DNA at the DSB as the transfected double-stranded sequences are used as templates for the repair enzymes.

3.4.3 Application of TALEN Technology

TALEN technology has been used for instance to efficiently engineer stably modified hESCs and iPSCs clones and human erythroid cell lines[44]. The technology has also been utilized experimentally to correct the genetic errors that underlie disease. For example, it has been used *in vitro* to correct the genetic defects that cause disorders such as sickle cell disease, xeroderma pigmentosum, and epidermolysis bullosa. It was also shown that TALEN technology can be used as tools to harness the immune system to fight cancers. In theory, the genome-wide specificity of engineered TALEN fusions allows for correction of errors at individual genetic loci via homology-directed repair from a correct exogenous template. In reality, however, in situ application of TALEN technology is currently limited by the lack of an efficient delivery mechanism, unknown immunogenic factors, and uncertainty in the specificity of TALEN binding. Another emerging application of TALEN technology is its ability to combine with other genome engineering tools, such as meganucleases. The DNA binding region of a TALE can be combined with the cleavage domain of a meganuclease to create a hybrid architecture combining the ease of engineering and highly specific DNA binding activity of a TALE with the low site frequency and specificity of a meganuclease.

In 2015, Physicians at the Great Ormond Street Hospital announced the first clinical use of TALEN-based genome editing. An 11-month-old baby suffering from CD19$^+$ acute lymphoblastic leukemia was treated with modified donor T cells that had been engineered to attack leukemia cells, to be resistant to Alemtuzumab, and to evade detection by the host immune system after introduction. A few weeks after therapy, the patient's condition improved; though physicians are cautious, the patient has been in remission for several months following treatment[45].

3.5 Genome Editing 3.0: CRISPR/Cas9 Tool

The CRISPR/Cas system is a prokaryotic immune system that confers resistance to foreign genetic elements such as plasmids and phages, and provides a form of acquired immunity[46]. CRISPR spacers recognize and cut these exogenous genetic elements in a manner analogous to RNA interference in eukaryotic organisms. A set of genes was found to be associated with CRISPR repeats, and was named the Cas, or CRISPR-associated, genes. The Cas genes encode putative nuclease or helicase proteins, which are enzymes that can cut or unwind DNA. The Cas genes are always located near the CRISPR sequences. There are a number of Cas enzymes, but the best known is called Cas9, which comes from streptococcus pyogenes.

The CRISPR interference technique has enormous potential application, including altering the germline of humans, animals and other organisms, and modifying the genes of food crops. By delivering the Cas9 protein and appropriate guide RNAs into a cell, the organism's genome can be cut at any desired location. CRISPRs have been used in concert with specific endonuclease enzymes for genome editing and gene regulation in species throughout the tree of life. Ethical concerns have been expressed about this nascent biotechnology and the prospect of editing the human germline.

3.5.1 CRISPR/Cas9 Mechanism

CRISPR/Cas9 genome editing is carried out with a Type Ⅱ CRISPR system. Cas9 is an enzyme (nuclease) that cuts DNA, and CRISPR is a collection of DNA sequences that tells Cas9 exactly where to cut. A guide RNA is required to feed Cas9 the right sequence, where to cut and paste bits of DNA sequence into the genome wherever you want. When utilized for genome editing, this system includes Cas9, CRISPR RNA (crRNA), trans-activating crRNA (tracrRNA) along with an optional section of DNA repair template that is utilized in either NHEJ or HDR. The crRNA contains the RNA used by Cas9 to guide it to the correct section of host DNA along with a region that binds to tracrRNA (generally in a hairpin loop form) forming an active complex with Cas9. The tracrRNA binds to crRNA and forms an active complex with Cas9.

CRISPR/Cas9 often employs a plasmid to transfect the target cells. The crRNA needs to be designed for each application as this is the sequence that Cas9 uses to identify and directly bind to the cell's DNA. The crRNA must bind only where editing is desired. The repair template must also be designed for each application, as it must overlap with the sequences on either side of the cut and code for the insertion sequence. Multiple crRNA and the tracrRNA can be packaged together to form a single-guide RNA (sgRNA). This sgRNA can be joined together with the Cas9 gene and made into a plasmid in order to be transfected into cells. The Cas9 protein with help of the crRNA finds the correct sequence in the host cell's DNA and creates a single or DSB in the DNA. Properly spaced single strand breaks in the host DNA can trigger homology directed repair, which is less error prone than NHEJ that typically follows a DSB. Providing a section of DNA repair template allows for the insertion of a specific DNA sequence at an exact location within the genome. The repair template should extend 40 to 90 base pairs beyond the Cas9 induced DNA break. The goal is for the cell's HDR process to utilize the provided repair template and thereby incorporate the new sequence into the genome. Once incorporated, this new sequence is now part of the cell's genetic material and passes into its daughter cells.

3.5.2 Application of CRISPR/Cas9 System

Like RNAi, CRISPR interference (CRISPRi) turns off genes in a reversible fashion by targeting, but not cutting a site. The targeted site is methylated so the gene is epigenetically modified. This modification inhibits transcription. Cas9 is an effective way of targeting and silencing specific genes at the DNA level. Cas9 was used to carry synthetic transcription factors (protein fragments that turn on genes) that activated specific human genes. CRISPR simplifies the creation of animals for research that mimic disease or show what happens when a gene is knocked down or mutated. CRISPR may be used at the germline level to create animals where the gene is changed everywhere. CRISPR can also be utilized to create human cellular models of disease. For instance, CRISPR was applied to human pluripotent stem cells to introduce targeted mutations in genes relevant to two different kidney diseases: polycystic kidney disease and focal segmental glomerulosclerosis[47].

3.6 How Does Genome Editing Work?

Genome editing uses a type of enzyme called an "engineered nuclease" which cuts the genome in a specific place. Engineered nucleases are made up of two parts: a nuclease part that cuts the DNA, a DNA-targeting part that is designed to guide the nuclease to a specific sequence of DNA, after cutting the DNA in a specific place, the cell will naturally repair the cut. We can manipulate this repair process to make changes (or "edits") to the DNA in that location in the genome.

3.7 Types of Genome Editing

3.7.1 Small DNA Changes

A nuclease enzyme is engineered to cut at a specific location in the DNA. After cutting the DNA with the engineered nuclease, the cell's normal DNA repair machinery will recognize the damage and join the two cut ends of DNA back together. This simple repair process is not 100 per cent perfect and often a few bases are lost or added around the site of the cut when it is repaired. This small change (mutation) in the DNA will affect the function of that section of DNA, which may mean a gene does not function properly or does not function at all.

3.7.2 Removal of a Section of DNA

To remove a section of DNA, nucleases are engineered that make cuts in the DNA either side of the section that we want to remove. After the engineered nucleases cut the DNA, the cell's normal DNA repair machinery will recognize the damage but may mistakenly join the wrong ends of DNA together, removing the DNA between the two cuts.

3.7.3 Insertion of a Section of DNA

A natural DNA repair system can be hijacked to insert a section of DNA into a genome by genome editing. Normally before a cell divides, all of its DNA is copied so that the two resulting daughter cells can receive a complete copy of the genome. If there is a break in one copy of the DNA, the cell repairs the break by using the other copy as a template. This process ensures that both copies of the DNA match again and is called "homology-directed repair".

It is possible with genome editing to take advantage of this DNA repair system to "trick" the cell into inserting a section of DNA. A nuclease enzyme is engineered to cut at a specific location in the DNA. After the DNA has been cut, a modified piece of DNA similar in sequence to the site of the cut is introduced. The cell uses the modified piece of DNA as the template to repair the break, filling the break with a copy of the new DNA.

This approach can be used to insert a new section of DNA, or to replace an existing section of DNA with an altered version, for example, to correct a point mutation within a gene.

3.8 The Future of Gene Therapy

(1) Human beings are transforming from solving previous infection-based diseases to chronic diseases with complex causes such as cancer, cardiovascular and cerebrovascular diseases, and neurodegenerative diseases.

(2) To solve these diseases, traditional drug interventions may sometimes be stretched, and a new "drug" is needed.

(3) Cell gene medicine is "upgraded" by integrating multiple cutting-edge biotechnologies such as antibody engineering, cell engineering, genetic engineering, and fermentation engineering into the cell platform. It is expected to become the next generation of drug intervention after small molecule drugs and macromolecular drugs.

(4) Understood the structure of genes sixty years ago, analyzed the sequence of genes through next-generation sequencing twenty years ago, and began to edit and modify their

genes skillfully ten years ago. Now that related products are successfully launched on the market and the rise of synthetic biology, more and more modified cellular gene drugs will gradually enter the clinic and even go on the market.

Supplement

List of Abbreviations	
AAV	adenovirus-associated virus
CAR-T	chimeric antigen receptor T cell
CRISPR	clustered regularly interspaced short palindromic repeats
CSNB	congenital stationary night blindness
HBV	hepatitis B virus
HCV	hepatitis C virus
IRDs	inherited retinal diseases
TALEN	transcription activator-like effector nuclease
TCR-T	T cell receptor-gene engineered T cells
ZFNs	zinc finger nucleases

Key Words List	
T 细胞受体基因工程改造的 T 细胞	TCR-T
规律间隔成簇短回文重复序列	CRISPR
嵌合抗原受体 T 细胞	CAR-T
锌指核酸酶	ZFNs
转录激活因子样效应物核酸酶	TALEN

References

[1] TAHER A T, WEATHERALL D J, CAPPELLINI M D. Thalassaemia [J]. Lancet, 2018, 391: 155 -167.
[2] NESTEROVA A P, et al. Disease pathways: an atlas of human disease signaling pathways [J]. Diseases of the blood, 2020, 95 -96.
[3] DISTELMAIER L, DÜHRSEN U, DICKERHOFF R. Sichelzellkrankheit [Sickle cell disease] [J]. Internist (Berl), 2020, 61: 754 -758.
[4] SINCLAIR A, ISLAM S, JONES S. Gene therapy: an overview of approved and pipeline technologies [J]. CADTH issues in emerging health technologies, 2018, 171: 1 -23.
[5] FRIEDMANN T, ROBLIN R. Gene therapy for human genetic disease? [J]. Science, 1972, 175: 949 -955.

[6] SHIMOTOHNO K, TEMIN H M. Formation of infectious progeny virus after insertion of herpes simplex thymidine kinase gene into DNA of an avian retrovirus [J]. Cell, 1981, 26: 67 -77.
[7] WEI C M, GIBSON M, SPEAR P G, et al. Construction and isolation of a transmissible retrovirus containing the src gene of Harvey murine sarcoma virus and the thymidine kinase gene of herpes simplex virus type 1 [J]. Journal of virology, 1981, 39: 935 -944.
[8] TABIN C J, HOFFMANN J W, GOFF S P, et al. Adaptation of a retrovirus as a eucaryotic vector transmitting the herpes simplex virus thymidine kinase gene [J]. Molecular and cellular biology, 1982, 2: 426 -436.
[9] BLAESE R M. Development of gene therapy for immunodeficiency: adenosine deaminase deficiency [J]. Pediatric research, 1993, 33: S49 - S55.
[10] RAPER S E, CHIRMULE N, LEE F S, et al. Fatal systemic inflammatory response syndrome in a ornithine transcarbamylase deficient patient following adenoviral gene transfer [J]. Molecular genetics and metabolism, 2003, 80: 148 - 158.
[11] HACEIN-BEY-ABINA S, VON KALLE C, SCHMIDT M, et al. LMO2-associated clonal T cell proliferation in two patients after gene therapy for SCID-X1 [J]. Science, 2003, 302: 415 -419.
[12] HACEIN-BEY-ABINA S, GARRIGUE A, WANG G P, et al. Insertional oncogenesis in 4 patients after retrovirus-mediated gene therapy of SCID-X1 [J]. Journal of clinical investigation, 2008, 118: 3132 -3142.
[13] CAVAZZANA-CALVO M, HACEIN-BEY S, DE SAINT BASILE G, et al. Gene therapy of human severe combined immunodeficiency (SCID) -X1 disease [J]. Science, 2000, 288: 669 -672.
[14] KASTELEIN J J, ROSS C J, HAYDEN M R. From mutation identification to therapy: discovery and origins of the first approved gene therapy in the Western world [J]. Human gene therapy, 2013, 24: 472 -478.
[15] SCHIMMER J, BREAZZANO S. Investor outlook: rising from the ashes; GSK's european approval of strimvelis for ADA-SCID [J]. Human gene therapy clinical development, 2016, 27: 57 -61.
[16] SILVERMAN E. Kymriah: a sign of more difficult decisions to come [J]. Managed care, 2018, 27: 17.
[17] Axicabtagene ciloleucel (Yescarta) for B-cell lymphoma [J]. Medical letter on drugs and therapeutics, 2018, 60: e122 - e123.
[18] DARROW J J. Luxturna: FDA documents reveal the value of a costly gene therapy [J]. Drug discovery today, 2019, 24: 949 -954.
[19] COX D B, PLATT R J, ZHANG F. Therapeutic genome editing: prospects and challenges [J]. Nature medicine, 2015, 21: 121 -131.

[20] SADELAIN M, BRENTJENS R, RIVIÈRE I. The basic principles of chimeric antigen receptor design [J]. Cancer discovery, 2013, 3: 388 - 398.

[21] SADELAIN M, RIVIÈRE I, BRENTJENS R. Targeting tumours with genetically enhanced T lymphocytes [J]. Nature reviews cancer, 2003, 3: 35 - 45.

[22] BAINBRIDGE J W, SMITH A J, BARKER S S, et al. Effect of gene therapy on visual function in Leber's congenital amaurosis [J]. New England journal of medicine, 2008, 358: 2231 - 2239.

[23] MAGUIRE A M, HIGH K A, AURICCHIO A, et al. Age-dependent effects of RPE65 gene therapy for Leber's congenital amaurosis: a phase 1 dose-escalation trial [J]. Lancet, 2009, 374: 1597 - 1605.

[24] HAUSWIRTH W W, ALEMAN T S, KAUSHAL S, et al. Treatment of leber congenital amaurosis due to RPE65 mutations by ocular subretinal injection of adeno- associated virus gene vector: short-term results of a phase I trial [J]. Human gene therapy, 2008, 19: 979 - 990.

[25] BUNTING S, ZHANG L, XIE L, et al. Gene therapy with BMN 270 results in therapeutic levels of FVIII in mice and primates and normalization of bleeding in hemophilic mice [J]. Molecular therapy, 2018, 26: 496 - 509.

[26] TALBOT K, TIZZANO E F. The clinical landscape for SMA in a new therapeutic era [J]. Gene therapy, 2017, 24: 529 - 533.

[27] MENDELL J R, AL-ZAIDY S, SHELL R, et al. Single-dose gene-replacement therapy for spinal muscular atrophy [J]. New England journal of medicine, 2017, 377: 1713 - 1722.

[28] MAEDER M L, GERSBACH C A. Genome-editing technologies for gene and cell therapy [J]. Molecular therapy, 2016, 24: 430 - 446.

[29] SHIM G, KIM D, PARK G T, et al. Therapeutic gene editing: delivery and regulatory perspectives [J]. Acta pharmacologica sinica, 2017, 38: 738 - 753.

[30] LI Q, QIN Z, WANG Q, et al. Applications of genome editing technology in animal disease modeling and gene therapy [J]. Computational and structural biotechnology journal, 2019, 17: 689 - 698.

[31] DOUDNA J A, CHARPENTIER E. Genome editing. The new frontier of genome engineering with CRISPR-Cas9 [J]. Science, 2014, 346: 1258096.

[32] RICROCH A. Global developments of genome editing in agriculture [J]. Transgenic research, 2019, 28: 45 - 52.

[33] URNOV F D, REBAR E J, HOLMES M C, et al. Genome editing with engineered zinc finger nucleases [J]. Nature reviews genetics, 2010, 11: 636 - 646.

[34] LIU Q, SEGAL D J, GHIARA J B, et al. Design of polydactyl zinc-finger proteins for unique addressing within complex genomes [J]. Proceedings of the National Academy

of Sciences of the United States of America, 1997, 94: 5525 - 5530.

[35] BITINAITE J, WAH D A, AGGARWAL A K, et al. FokI dimerization is required for DNA cleavage [J]. Proceedings of the National Academy of Sciences of the United States of America, 1998, 95: 10570 - 10575.

[36] PEREZ E E, WANG J, MILLER J C, et al. Establishment of HIV-1 resistance in CD4 + T cells by genome editing using zinc-finger nucleases [J]. Nature biotechnology, 2008, 26: 808 - 816.

[37] HOLT N, WANG J, KIM K, et al. Human hematopoietic stem/progenitor cells modified by zinc-finger nucleases targeted to CCR5 control HIV-1 in vivo [J]. Nature biotechnology, 2010, 28: 839 - 847.

[38] DIDIGU C A, WILEN C B, WANG J, et al. Simultaneous zinc-finger nuclease editing of the HIV coreceptors ccr5 and cxcr4 protects CD4 + T cells from HIV-1 infection [J]. Blood, 2014, 123: 61 - 69.

[39] JOUNG J K, SANDER J D. TALENs: a widely applicable technology for targeted genome editing [J]. Nature reviews molecular cell biology, 2013, 14: 49 - 55.

[40] GAJ T, GERSBACH C A, BARBAS C F. ZFN, TALEN, and CRISPR/Cas-based methods for genome engineering [J]. Trends in biotechnology, 2013, 31: 397 - 405.

[41] MAK A N, BRADLEY P, CERNADAS R A, et al. The crystal structure of TAL effector PthXo1 bound to its DNA target [J]. Science, 2012, 335: 716 - 719.

[42] DENG D, YAN C, PAN X, et al. Structural basis for sequence-specific recognition of DNA by TAL effectors [J]. Science, 2012, 335: 720 - 723.

[43] SYMINGTON L S. Mechanism and regulation of DNA end resection in eukaryotes [J]. Critical reviews in biochemistry and molecular biology, 2016, 51: 195 - 212.

[44] HOCKEMEYER D, WANG H, KIANI S, et al. Genetic engineering of human pluripotent cells using TALE nucleases [J]. Nature biotechnology, 2011, 29: 731 - 734.

[45] POIROT L, PHILIP B, SCHIFFER-MANNIOUI C, et al. Multiplex genome-edited T-cell manufacturing platform for "off-the-shelf" adoptive T-cell immunotherapies [J]. Cancer research, 2015, 75: 3853 - 3864.

[46] WIEDENHEFT B, STERNBERG S H, DOUDNA J A. RNA-guided genetic silencing systems in bacteria and archaea [J]. Nature, 2012, 482: 331 - 338.

[47] BEN JEHUDA R, SHEMER Y, BINAH O. Genome editing in induced pluripotent stem cells using CRISPR/Cas9 [J]. Stem cell reviews and reports, 2018, 14: 323 - 336.